Lecture Notes in Control and Information Sciences 238

Editor: M. Thoma

Springer-Verlag London Ltd.

Nicola Elia and Munther A. Dahleh

Computational Methods for Controller Design

Springer

Authors

Nicola Elia
Munther A. Dahleh
Laboratory for Information and Decision Systems, Massachusetts Institute of Technology, 127 Massachusetts Avenue, Cambridge, MA 02139, USA

ISBN 978-1-85233-075-0

British Library Cataloguing in Publication Data
Elia, Nicola
Computational methods for controller design. - (lecture notes in control and information sciences ; 238)
1. Automatic control - Data processing
I. Title II. Dahleh, Munther A.
629.8'312'0285
ISBN 978-1-85233-075-0

Library of Congress Cataloging-in-Publication Data
Elia, Nicola, 1962 -
Computational methods for controller design/ Nicola Elia and Munther A. Dahleh.
p. cm. -- (Lecture notes in control and information sciences ; 238)
ISBN 978-1-85233-075-0 ISBN 978-1-84628-532-5 (eBook)
DOI 10.1007/978-1-84628-532-5
1. Automatic control -- Mathematical models. 2. Numerical analysis.
3. Linear programming. I. Dahleh, Munther A. II. Title
III. Series.
TJ213. E562 1998 98-34634
629.8--dc21 CIP

Originally published by Springer-Verlag London Berlin Heidelberg in 1998

Typesetting: Camera ready by authors

69/3830-543210 Printed on acid-free paper

Preface

The monograph describes a computationally-based methodology for controller design that can handle typical time and frequency domain specifications and provide a characterization of the limits of performance for a given system. The methodology is based on optimization, where one performance objective is optimized and the other specifications are the constraints. Many practical performance objectives can be represented in terms of convex constraints on the closed loop system response. Such objectives include standard norm constraints, bandwidth constraints, time templates on the closed loop response, as well as stability robustness conditions for unstructured uncertainty. With this, the design problem is turned into an infinite dimensional convex optimization problem. The infinite dimension arises from the fact that the set of feasible closed loop maps (corresponding to a stabilizing controller) is infinite.

A key issue in this book is understanding how do we solve these problems. The solvability of a problem describes when and in what sense the solution can be computed with reasonable complexity.

While finite dimensional convex problems can be efficiently solved, the situation is more complicated with infinite dimensional ones. There is a large class of problems that have an underlying finite dimensional structure. It is interesting to note that in the $\mathcal{H}_\infty$ and ℓ_1 methodologies, one-block problems belonged to this class. Characterizing such classes is an important component of this research.

When the problem is truly infinite dimensional, or when no finite dimensional structure can be revealed, the solvability of the problem is characterized by the existence of computable approximate solutions, the accuracy of the approximations, and by the information provided by the approximate solutions about the structure of the optimal solution or the optimal controller.

It turns out that duality theory is a fundamental tool in analyzing the solvability of infinite dimensional generalized linear programs, and in providing generic computational methods for such problems.

In this book, we develop a uniform treatment of multi-objective control problems by providing

- a unified way to pose the problems as generalized linear programs and to derive their duals,
- duality theory results that characterize the duality relationship for the generalized linear programs arising from multi-objective control problems,

- a set of tools to analyze the convergence properties of the computational method based on the duality relationship,
- the complete analysis and extension of methods developed for the ℓ_1 problem, for several important multi-objective problems which makes them readily implementable and usable for design.

In summary, the book provides the reader with a rather complete guide on how pose practical multi-objective control problems in this framework, and how to solve them, i.e., how to derive and analyze readily implementable computational methods to derive exact or approximate solutions.

Organization

The monograph is organized in ten chapters. Chapter 1 contains the introduction to the material in the book, and gives an overview on the area of computational approaches to controller design. Chapter 2 contains most of the necessary notation and the mathematical preliminaries that are needed in the development of the investigation. Chapter 3 contains the control problem setup, the description of the stability constraints and of several typical performance objectives, and shows how these specifications are equivalent to generalized linear constraints. In Chapter 4, the duality theory results for the generalized linear programming problems are derived. These results are applied in the following three chapters to the analysis of several multi-objective problems. Chapter 5 contains the complete treatment for the ℓ_1 control problem with time-domain constraints on the response to fixed inputs. Two cases are considered: 1) the constraints are imposed only for finite time (finite horizon case). 2) The constraints are imposed for infinite time (infinite horizon case). Chapter 6 contains the solvability analysis of the ℓ_1 optimal control problem with frequency point magnitude constraints. This problem is a convex optimization with infinite dimensional Linear Matrix Inequality constraints. The main point in this chapter is that the fact that primal and dual problem have the same cost (no of duality gap) in an infinite dimensional problem does not imply that we can compute converging primal and dual finite dimensional approximations. Chapter 7 analyzes the mixed $\mathcal{H}_2/\ell_1$ problem. The derivations for the mixed $\ell_1/\mathcal{H}_2$ problem are analogous and are omitted. In Chapter 8, a new computational method for ℓ_1 is presented and its properties are analyzed. The issues of deriving exact or approximate solutions are similar when the problems are posed as dynamic games in state space instead of as convex optimizations on the space of the closed loop maps. Chapter 9 contains a dynamic programming derivation of the optimal (nonlinear) static full state feedback strategy that minimizes the worst-case peak-to-peak gain of the closed loop system. Both finite and infinite horizon problems are considered. Once again, that duality theory provides important extra information about the problem and allows the derivation of the structure of the optimal strategy and of approximation methods when the optimal strategy cannot be computed exactly. Finally, Chapter 10 presents the conclusions.

Acknowledgments

We are grateful to Petros Voulgaris, Fernando Paganini, Saligrama Venkatesh, Michael Branicky, Mitch Livstone, and Jeff Shamma. They all have influenced the shaping of this work in various ways. In particular, we would like to thank Sanjoy Mitter, Alexander Metgreski, and Peter Young for being invaluable sources of discussions, suggestions, and encouragement, the sponsors NSF, Draper Laboratories, AFOSR, and ARO for their support, and our spouses for their constant dedication and understanding.

Cambridge, MA, *Nicola Elia*

Munther A. Dahleh

Contents

1. Introduction

1.1 Background and Motivation

Many of the current control applications involve multi-input multi-output complex models for which a controller is needed to deliver tight performance specifications in the presence of plant uncertainty, input uncertainty, and actuator constraints. Since classical control methods cannot deal with such problems, several methodologies emerged in the past two decades that systematically address the fundamental limitations and capabilities of linear feedback controllers for MIMO uncertain systems. Examples of such methodologies are the $\mathcal{H}_2$, $\mathcal{H}_\infty$, and ℓ_1 approaches.

The $\mathcal{H}_2$ (LQG) design method is based on minimizing the RMS value of an error signal in the presence of white noise exogenous input. The $\mathcal{H}_\infty$ design method is based on minimizing the maximum energy of an error signal in the presence of finite energy unknown exogenous inputs. And, in a similar fashion, the ℓ_1 problem is based on minimizing the maximum peak value of an error signal in the presence of finite amplitude unknown exogenous inputs. The problem of finding a controller that minimizes any of the above measures for a nominal plant is computationally tractable and software for designing such controllers is readily available for practitioners. Plant uncertainty is more naturally described in terms of the $\mathcal{H}_\infty$ or ℓ_1 norms, and thus when present, these methods can be extended easily to account for the stability robustness and performance robustness of the system. The resulting design problems in these cases involve scaled $\mathcal{H}_\infty$ or ℓ_1 problems (known as μ analysis and synthesis in the case of $\mathcal{H}_\infty$). Even though these problems are nonconvex, iterative methods for providing good suboptimal controllers are readily available. A complete account of these methodologies can be found in [1, 2, 3]. Recent work has focused on extending the $\mathcal{H}_2$ approach to handle plant uncertainty utilizing the power of Integral-Quadratic-Constraints as a general framework for robustness analysis and synthesis [4, 5, 6].

While these methodologies are powerful, they still fall short from giving a complete characterization of the limits of performance for a given system. Many practical applications require several types of specifications that cannot be addressed by one of the above methodologies. In particular, a combination of time-domain and frequency-domain specifications is quite natural in one application; e.g., tracking with bandwidth constraints. As a consequence, computationally-based methodologies emerged to address problems of this nature. This book addresses one such methodology.

1.2 Multi-Objective Control as Convex Optimization

The limitations of the methodologies described above can be summarized as follows:

- none can handle combined time and frequency domain specifications,
- none can address specifications concerning a fixed input, and
- all are norm-based methods.

It is well known, however, that many of the design specifications on the closed loop system are convex [7, 1]. This has motivated the development of optimization based approaches that can address the general design problem. Examples of such approaches are Q-Design [7], the use of linear matrix inequalities (LMI) [8], and generalized linear programming which is the topic of this book.

In [7, 1], a catalog of interesting performance specifications that translate into convex constraints on the closed loop map is presented. Such objectives include standard norm constraints, bandwidth constraints, time templates on the closed loop response, as well as stability robustness conditions for unstructured uncertainty. With this, the design problem is turned into an infinite dimensional convex optimization problem. The infinite dimension arises from the fact that the set of feasible closed loop maps (corresponding to a stabilizing controller) is infinite. Understanding how to solve such problems is clearly the fundamental issue in the development of this general methodology.

It is interesting to note that there are some special cases where different objectives can be well represented with different norms. These cases have motivated the important multi-objective control problems based on mixed norm minimization considered in [9, 10, 11, 12, 13]. Apart from their limited applicability, the main drawback of mixed norm minimization problems is that each has a different solution method. The lack of a unifying theme is a prime motivation behind the development in this book.

1.3 Solutions of Infinite Dimensional Convex Problems

While finite dimensional convex problems can be efficiently solved, the situation is more complicated with infinite dimensional ones. However, there is a large class of problems that have an underlying finite dimensional structure. Characterizing such classes is an important component of this research. It is interesting to note that in the $\mathcal{H}_\infty$ and ℓ_1 methodologies, one-block problems belonged to this class.

When the problem is truly infinite dimensional, or when no finite dimensional structure can be revealed, the solvability of the problem is characterized by:

- Existence of computable approximate solutions.
- Accuracy of the approximations.
- Information provided by the approximate solutions about the structure of the optimal solution or the optimal controller.

It turns out that duality theory is a fundamental tool in analyzing the solvability of infinite dimensional convex problems.

1.3.1 Duality Theory and Generalized Linear Programs

In this book, we develop a uniform treatment of multi-objective control problems by transforming them into generalized linear programming problems. It is interesting to note that most convex optimization problems can be represented as linear programming problems, with equality constraints, and an appropriate positive cone. These linear programs can be shown to have special structures that enables us to provide either exact or approximate solutions.

The main tool used in the analysis of the generalized linear programs is Duality theory. Duality theory is a mathematical tool that provides alternative (dual) optimization problems that are related to the original (primal) optimization problems. For finite dimensional linear programs, duality theory is well developed and has been successfully exploited to provide efficient algorithms for obtaining solutions. On the other hand, duality theory for infinite dimensional linear programs is more complicated, and although many results exist in this area, we present some useful results that exploit certain common features arising from multi-objective control problems.

A contribution of this monograph is to characterize, in a precise way, the duality relationships between the primal and dual problems in this specific class of infinite dimensional linear programs. Such relationships are not as straightforward as they are in the finite dimensional case. In particular, a given primal problem admits more than one dual representation and has different relations with these duals. The weakest of such relations involves cases where the optimal cost of the dual problem is always smaller than the optimal cost of the primal. The strongest involves cases where the primal and dual problems have the same optimal cost (no duality gap), the optimal solutions of both primal and dual problems exist and are related to each other in a precise way. The stronger is the duality relationship, the more the primal-dual pair provides insight into the structure of the problem and its solution.

The characterization of the duality relationship is only the first step in the analysis of the optimization problem. In some cases, a finite dimensional structure may become apparent in the primal or the dual problem, suggesting ways to compute the solution exactly. In general, however, it is important that the infinite dimensional primal and/or dual problem can be approximated by finite dimensional (computable) optimization problems. One can argue that it is always possible to find a dual to every linear program without a duality gap, however, it may not be possible to find one with the right approximation properties.

1.3.2 Computational Methods

One of the main objectives of this research is to study general methods to approximate a given infinite dimensional optimization problem using finite dimensional problems. Solutions to finite dimensional convex optimization problems can be computed relatively easily. Several efficient algorithms are now available even for non-differentiable optimization costs [7, 8]. The fundamental issue, however, is the quality of the approximation.

There are many criteria to evaluate the quality of an approximation method. The minimum requirement is that the approximate cost approaches the optimal cost of the infinite dimensional problem as the order of the approximation increases. Many methods presented in the literature possess only this property. An additional important requirement is that the method provides an accuracy estimate on how close the approximation is to the optimal cost. Most approximation methods possessing this property provide upper and lower bounds to the optimal cost. It is also desirable that both upper and lower bounds converge to the optimal cost as the order of the approximation increases. With such methods, the optimal cost can be computed within any desired accuracy. In some cases, the accuracy can be computed a priori and is a function of the approximation order only.

In addition to the accuracy, a good approximation method also provides approximate suboptimal solutions that converge to the optimal solution. These methods are also classified according to the kind of convergence they provide. Convergence on compact sets is usual, norm convergence is less common, and convergence in a finite number of steps is very desirable, but extremely rare.

In this monograph, we consider general computational methods based on approximating the primal and dual problems with finite support feasible solutions. The primal approximation provides suboptimal feasible solutions to the problem, while the dual approximation provides lower bounds of the dual cost which translate to bounds on the accuracy of the primal approximation. These primal and dual approximation methods are known in the ℓ_1 literature as the Finitely Many Variables method (FMV) [14, 15, 1], and the Finitely Many Equations method (FME) [16, 15, 1]. The extension of these methods to multi-objective control problems is analyzed, and their properties are derived.

Other methods based on duality can be analyzed or proposed. They differ only in the way feasible primal and dual solutions are selected. For example, similar analysis can be done for the Delay Augmentation method [16, 1].

1.3.3 Applications

Several applications of the general approach will be presented. These are samples of more general multi-objective problems that can be studied or solved by the results presented in this monograph.

The ℓ_1 Control Problem with Time-Domain Constraints. In this problem, we search for a stabilizing controller that satisfies the following two objectives: minimize the ℓ_1 norm of some closed loop transfer function and guarantee that the output signal due to a fixed input is inside a given template for all times. This problem is important in practice, since control engineers define many design specifications in terms of the response of the feedback system to particular fixed inputs. Particular attention is given to problems with time domain specifications on the step response such as overshoot, undershoot, settling time, which are commonly used in practice.

The ℓ_1 Optimal Control Problem with Magnitude Constraints in the Frequency Domain. This is the problem of finding a stabilizing controller such that the ℓ_1 norm of the closed loop transfer function is minimized subject to constraints imposed on the magnitude of its frequency response. In particular, we study the solvability of the problem where the frequency constraints are imposed at a finite number of frequency points. This problem helps us in making the point that having a primal-dual formulation with no duality gap does not automatically imply that we can compute converging primal and dual finite dimensional approximations.

The Mixed $\mathcal{H}_2/\ell_1$ Control Problem. The $\mathcal{H}_2$ optimal control problem remains the most popular multivariable controller design methodology for Multi-Input Multi-Output systems. There are several interpretations of the meaning of this optimization problem. In the most common interpretation, the optimal $\mathcal{H}_2$ closed loop system provides the minimum variance of the outputs in presence of white noise disturbances. The $\mathcal{H}_2$ design method has an elegant closed-form solution, the optimal controller has interesting separation properties, and the method has a rich history of successful applications. Nevertheless, solutions do not have guaranteed stability margins against model uncertainty. This limitation has motivated much research toward methodologies that can directly incorporate model uncertainty, such as $\mathcal{H}_\infty$ and ℓ_1. Solving the mixed $\mathcal{H}_2/\ell_1$ problem allows the design of systems that optimally reject white noise input and that, at the same time, have a guaranteed stability margin against model uncertainty.

1.3.4 A New Method for ℓ_1

The ability to analyze and solve multi-objective control problems has also allowed us to derive a new computational method for the ℓ_1 problem that overcomes many of the drawbacks that the other methods have. The method is based on solving a sequence of special mixed $\ell_1/\mathcal{H}_2$ problems, each of which results in a finite dimensional convex optimization that can be solved exactly. The sequence of solutions to such problems converges to the optimal solution of the ℓ_1 problem.

1.3.5 Computational Methods in State Space

Although most of this monograph is concerned with multi-objective control problems as convex optimizations on the space of the closed loop maps, the issues of deriving exact or approximate solutions are similar when the problems are posed as dynamic games in state space. Multi-objective control problems are usually difficult to pose in state space, since many performance objectives are naturally imposed on the closed loop system and are difficult to transform into constraints on the system's states. Although state space approaches for multi-objective control problems are much less developed, solutions to these problems can provide additional and valuable information about the structure of the optimal strategies.

In this monograph, we consider the problem of finding the state feedback controller that minimizes the worst-case peak-to-peak amplification of the closed loop system. If we search over the class of linear controllers, this becomes the standard ℓ_1 problem. Recent results [17, 18, 19] have shown that nonlinear static controllers can provide strictly better performance than the linear, possibly dynamic, controllers. We propose a state space approach to the solution of this problem, based on dynamic programming and duality theory. The proposed approach encompasses as a special case the method presented in [19] based on viability theory. Moreover, it is directly generalizable to the mixed objective problem where also hard actuator saturation constraints must not be violated at all times.

Once again, duality theory provides important extra information about the problem and allows the derivation of the structure of the optimal strategy and of approximation methods when the optimal strategy cannot be computed exactly.

1.4 Main Contributions

The main contributions are now summarized.

- Proposed a unified approach, based on generalized linear programs, to solve convex optimization problems arising in multi-objective control problems.
- Developed general duality theory results for this class of problems.
- Extended computational methods developed for ℓ_1 to multi-objective control problems.
- Shown the effectiveness of the approach in analyzing the solvability of several multi-objective control problems and in uncovering the finite dimensional structure of one-block problems for the examples considered.
- Proposed a new computational method for the ℓ_1 problem, which is superior to existing approaches and is based on the solution of a mixed objective problem. The main advantages of this method are that

  - the (sub-)optimal controller is easily computable
  - the convergence in norm to the optimal solution is guaranteed
  - the reordering of inputs and outputs (necessary in the DA method [16]) is no longer needed.
- Derived the optimal state feedback controller that minimizes the peak-to peak gain (finite horizon case).
  - Established the existence of a (nonlinear) time-invariant stabilizing controller in the infinite horizon case.
  - Derived suboptimal (nonlinear) time-invariant stabilizing controllers for the infinite horizon problem.

2. Mathematical Preliminaries

In this chapter, we establish the notation that will be used throughout the monograph. Apart from some minor differences, we follow quite closely the notation in [1]. The goal of this chapter is not to provide a complete treatment of the mathematical background, but only to establish a standard terminology and notation. The reader can find a more through treatment of certain topics in the reference material.

2.1 Normed Spaces

All the vector spaces we consider are Banach spaces, i.e., complete, normed vector spaces.

The *norm* of an element x belonging to the vector space X is indicated as $\|\cdot\|$.

A *linear functional* f on the space X is a linear mapping from X to the reals $\mathbb{R}$.

f is *bounded* if it has bounded gain, i.e., if

$$\|f\| \triangleq \sup_{x\in X, x\neq 0} \frac{|f(x)|}{\|x\|} < \infty \tag{2.1}$$

The space of all bounded linear functional on X, is denoted by X^* and called the *dual space* of X. X^*, with the norm defined by Equation (2.1), is also a Banach space. We will use the notation $\langle x, f\rangle$ to denote $f(x)$.

Examples of Banach Spaces. We introduce the definition of some Banach spaces that will arise in the treatment ahead.

Definition 2.1.1. *$\ell_1^{m\times n}$ is the space of all sequences, H, of $m\times n$ real matrices such that*

$$\|H\|_1 \triangleq \max_{1\leq i\leq m} \sum_{j=1}^{n} \|h_{ij}\|_1 < \infty,$$

where $\|h\|_1 \triangleq \sum_{t=0}^{\infty} |h_{ij}(t)|$.

We denote the space $\ell_1^{1\times 1}$ by simply ℓ_1. The notation ℓ_∞ and $\ell_\infty^{m\times n}$ denotes the dual spaces of ℓ_1 and $\ell_1^{m\times n}$. These spaces are characterized as follows:

Definition 2.1.2. *$\ell_\infty^{m\times n}$ is the space of all sequences G of $m\times n$ real matrices such that*

$$\|G\|_\infty \triangleq \sum_{i=1}^{m} \max_{1\le j\le n} \|g_{ij}\|_\infty < \infty,$$

where $\|g\|_\infty \triangleq \max_{0\le t<\infty} |g(t)|$.

The action of every element $G \in \ell_\infty^{m\times n}$ on any element $H \in \ell_1^{m\times n}$ can be represented uniquely in the form:

$$\langle H, G\rangle = \sum_{i=1}^{m}\sum_{j=1}^{n}\sum_{t=0}^{\infty} g_{ij}(t)h_{ij}(t).$$

The subspace of $\ell_\infty^{m\times n}$ consisting of all the bounded sequences of $m\times n$ real matrices for which

$$\lim_{t\to\infty} g_{ij}(t) = 0 \qquad \forall\, i = 1,\ldots,m, \qquad \forall j = 1,\ldots,n$$

is denoted by $c_0^{m\times n}$. The dual of $c_0^{m\times n}$ is $\ell_1^{m\times n}$ ($(c_0^{m\times n})^* = \ell_1^{m\times n}$).

Definition 2.1.3. *$\ell_2^{m\times n}(\mathbf{Z})$ denotes the Hilbert space of sequences of complex-valued $m\times n$ matrices, with inner product defined as follows.*

$$\langle H, G\rangle = \sum_{k=-\infty}^{\infty} Trace(\bar{G}(k)^T H(k)).$$

$\ell_2^{m\times n}(\mathbf{Z})$ can be written as the direct sum of two spaces of one-sided sequences

$$\ell_2^{m\times n}(\mathbf{Z}_+) \oplus \ell_2^{m\times n}(\mathbf{Z}_-),$$

with $0 \in \mathbf{Z}_+$.

The Fourier transform of G in $\ell_2^{m\times n}(\mathbf{Z})$ is defined as

$$\hat{G}(e^{-i\theta}) = \sum_{k=-\infty}^{\infty} G(k)e^{-ik\theta}$$

$\mathcal{L}_2^{m\times n}[0,2\pi)$ denotes the space whose elements are the Fourier Transform of elements in $\ell_2^{m\times n}(\mathbf{Z})$. The decomposition of $\ell_2^{m\times n}(\mathbf{Z})$ into $\ell_2^{m\times n}(\mathbf{Z}_+)$ and $\ell_2^{m\times n}(\mathbf{Z}_-)$ induces, through the Fourier Transform, an orthogonal decomposition of $\mathcal{L}_2^{m\times n}[0,2\pi)$:

$$\mathcal{L}_2^{m\times n}[0,2\pi) = \mathcal{H}_2^{m\times n} \oplus \mathcal{H}_2^{m\times n\perp},$$

where $\mathcal{H}_2^{m\times n}$ contains all the matrix-valued functions in $\mathcal{L}_2^{m\times n}[0,2\pi)$ that are analytic inside the open unit disc, and $\mathcal{H}_2^{m\times n\perp}$ contains all the matrix-valued functions in $\mathcal{L}_2^{m\times n}[0,2\pi)$ analytic in the complement of the unit disc. $\mathcal{RH}_2^{m\times n}$ is the space of the rational transfer function matrices in $\mathcal{H}_2^{m\times n}$. $\mathcal{RH}_2^{m\times n\perp}$ is defined analogously.

Since we will mostly work with real-valued unilateral matrix sequences supported on the positive integers, for notational convenience, we will denote $\ell_2^{m\times n}(\mathbf{Z}_+)$ as $\ell_2^{m\times n}$. For notational simplicity, we will often drop the superscripts $m\times n$ when no confusion arises.

2.1.1 LMI Spaces

In this section, we follow [20, 21], and introduce a Banach space of symmetric, possibly infinite, matrices. This space is useful in describing an important class of convex constraints, known as Linear Matrix Inequality Constraints, that will be introduced later on.

Let $\mathbb{R}_S^{m\times m}$ denote the set of symmetric matrices in $\mathbb{R}^{m\times m}$. The notion of symmetric matrix can be extended to infinite matrices as follows:

Definition 2.1.4. *Associate, to any two dimensional sequence $\{S_{ij}\}_{i,j=0}^{\infty}$, an infinite matrix S whose $(i,j)^{th}$ element is S_{ij}. $S=[S_{ij}]$. An infinite matrix $S=[S_{ij}]$ is symmetric if $S_{ij}=S_{ji}$ for all $i,j=0,\ldots,\infty$. $\mathbb{R}_S^{\infty}$ denotes the space of all (square) infinite symmetric matrices.*

For square, possibly, infinite matrices, we denote the induced $\ell_2-\ell_2$ norm by

$$\|S\|_{2i} \triangleq \sup_{v\neq 0}\frac{\|Sv\|_2}{\|v\|_2}$$

and we define

$$\ell_{2i}=\{S:\|S\|_{2i}<\infty\}.$$

For a finite matrix we have $\|S\|_{2i}=\bar{\sigma}(S)$, where $\bar{\sigma}$ denotes maximum singular value.

Definition 2.1.5. *An LMI space is a linear vector space whose elements, S, are square symmetric matrices. For infinite-size matrices we require that*

$$v^TSv \triangleq \sum_{i,j=1}^{\infty} S_{ij}v_iv_j \;\triangleq\; \lim_{m\to\infty}\left(\sum_{i,j=1}^{m} S_{ij}v_iv_j\right)$$

converges and $v^TSv<\infty$ for all $v\in\ell_2$.

Theorem 2.1.1. *The set*

$$\ell_{2i}^{LMI} \triangleq \{S : S \text{ is square symmetric and } S\in\ell_{2i}\}$$

is an LMI space, and, with the norm $\|S\|\triangleq\|S\|_{2i}$ is an LMI Banach space i.e., a complete normed LMI space.

Remark 2.1.1. For finite matrices, $\mathbb{R}_S^{m\times m}$ is a finite-dimensional vector space equivalent to $\mathbb{R}^{\frac{m(m+1)}{2}}$. Since all norms will induce the same topology on the space, it is more convenient to equip $\mathbb{R}_S^{m\times m}$ with the Frobenious norm (instead of ℓ_{2i}) defined as $\|S\|_F \triangleq \sqrt{Trace(S^TS)}$. Since any linear functional on $\mathbb{R}_S^{m\times m}$ can be represented as follows

$$\langle \Gamma, S\rangle = \sqrt{Trace(\Gamma S)} \qquad \text{with } \Gamma \in \mathbb{R}_S^{m\times m}$$

Then, it follows that $(\mathbb{R}_S^{m\times m}, \|\cdot\|_F)^* = (\mathbb{R}_S^{m\times m}, \|\cdot\|_F)$.

2.2 Convex Cones

By introducing a cone defining the positive vectors in a given space, it is possible to define a partial order relationship between elements of the space. We first recall the definition of a convex set.

Definition 2.2.1. *A set S in a linear vector space X is* convex *if for any $x_1, x_2 \in S$, all points of the form $s = \alpha x_1 + (1-\alpha)x_2$ with $0 \leq \alpha \leq 1$ belong to S.*

Definition 2.2.2. *A subset W of a real topological vector space is called a* cone *if it satisfies the following conditions:*

1) If $w \in W$ then $\alpha w \in W$ for all $\alpha \geq 0$
2) If $w_1, w_2 \in W$ then $w_1 + w_2 \in W$

Notice that, as it results from the above definition, a cone is automatically convex.

Definition 2.2.3. *[22] Let P be a cone in X. For $x,\ y \in X$, we write $x \geq y$ (with respect to P), if $x - y \in P$. The cone defining this relation is called the positive cone in X.*

In most situations, the choice of P will arise naturally. For example, if $X = \mathbb{R}^n$, it is natural to say that $x \geq y$ if all the components of $x - y$ are greater or equal to zero. In the space of the continuous functions defined on an interval $[a, b]$ of the real line, it is natural to define the positive cone as the set of all continuous functions in the space that are nonnegative everywhere on $[a, b]$. Nevertheless, the choice of the positive cone P can be quite arbitrary. An appropriate choice of P allows us to represent most convex optimization problems as abstract linear programming problems.

Definition 2.2.4. *For a subset $W \subset X$ of a Banach space X, set:*

$$W^{\oplus} = \{x^* \in X^* : \langle x, x^*\rangle \geq 0, \quad \forall x \in W\}$$

$$W^{\perp} = \{x^* \in X^* : \langle x, x^*\rangle = 0 \quad \forall x \in W\}$$

Notice that, for any $W \in X$, $W^{\oplus}$ is a closed cone in X^* and $W^{\perp}$ is a closed subspace of X^*. $W^{\oplus}$ is called the *positive conjugate cone* of W and $W^{\perp}$ is called the annihilator of W.

2.3 Convergence of Sequences

Now, we recall some notions of convergence useful in the development of the theory presented in this monograph.

Definition 2.3.1. *A sequence* $\{x_n\}$ *in a linear normed vector space* X *is said to* converge in norm *(or strongly) to* $x \in X$ $(x_n \to x)$, *if* $\|x - x_n\| \to 0$.

Definition 2.3.2. *A sequence* $\{x_n^*\}$ *in a linear normed vector space* X^* *is said to* converge *weak* to* $x^* \in X^*$ $(x_n^* \stackrel{w^*}{\to} x^*)$, *if, for any* $x \in X$, $\langle x, x_n^* \rangle \to \langle x, x^* \rangle$.

Definition 2.3.3. *A set* $\mathcal{S} \subset X^*$ *is weak** compact *if, for any sequence in* $\mathcal{S}$, *there exists a subsequence which is weak* convergent.*

Although a *weak** closed set $\mathcal{S}$ contains all its *weak** limit points, i.e., the limits of *weak** convergent sequences, the *weak** closure of $\mathcal{S}$ may contain points that are not *weak** limit points of $\mathcal{S}$. However, if $\mathcal{S}$ is a convex set in a Banach space X dual of a separable Banach space, then, the notions of *weak** closure and *weak** sequential closure coincide.

Proposition 2.3.1. *([23] Cor. V.12.7) A convex set in a Banach space* X *dual of a separable Banach space, is weak* closed if and only if it is weak* sequentially closed, i.e., if and only if it contains all its weak* limit points.*

Thus, in a Banach space dual of a separable Banach space, to check that a convex set is *weak** closed, it is enough to verify that it contains all the elements which are the limits of *weak** convergent sequences. This result will be implicitly used throughout the monograph since the conditions for its validity will always be satisfied. For example, the result will hold for any convex set in $\ell_1^{m \times n}$ since it is the dual of $c_0^{m \times n}$ which is separable.

Finally we conclude this section recalling the important Alaoglu's theorem.

Theorem 2.3.1. *[23] If* X *is a normed space, then, the closed unit ball* $\{x^* \in X^* \mid \|x^*\| \leq 1\}$ *in* X^* *is weak* compact.*

2.4 Bounded Linear Operators

Definition 2.4.1. T *is said to be a linear operator from space* X *to space* Y, *if satisfies*

$$T(\alpha x_1 + \beta x_2) = \alpha T x_1 + \beta T x_2 \qquad \forall \alpha, \beta \in \mathbb{R}$$

for each $x_1, x_2 \in X$.

Definition 2.4.2. *An operator* T *mapping the space* X *into the space* Y *is bounded if*

$$\sup_{x \neq 0} \frac{\|Tx\|}{\|x\|} < \infty.$$

The induced norm of the operator T is denoted by $\|T\|$ and is given by:

$$\|T\| = \sup_{x \neq 0} \frac{\|Tx\|}{\|x\|}.$$

In other words, the induced norm provides a measure of the worst-case gain of the operator. The following properties of bounded linear operators will be useful later on [24].

Theorem 2.4.1. *Let T be a linear operator from X to Y, then*

1. *T is continuous if and only if it is bounded.*
2. *If T is bounded, then the null space of T, $\mathcal{N}(T)$, is closed.*

$\mathcal{B}(X, Y)$ denotes the set of all bounded linear operators from the Banach space X to the Banach space Y.

Definition 2.4.3. *Let X and Z be normed spaces and let $\mathcal{T} \in \mathcal{B}(X, Z)$. The* adjoint *operator $\mathcal{T}^* : Z^* \to X^*$ is defined by the following equation:*

$$\langle x, \mathcal{T}^* z^* \rangle = \langle \mathcal{T} x, z^* \rangle$$

Given $\mathcal{T}$, a bounded linear operator from Z^* to X^*, ${}^*\mathcal{T} : X \to Z$ denotes the pre-adjoint, *when it exists, of $\mathcal{T}$, i.e., $({}^*\mathcal{T})^* = \mathcal{T}$.*

Theorem 2.4.2. *[24] Let T_n be a sequence of bounded linear operators from X to Y, such that $T_n x$ converges in norm to Tx in Y, for every $x \in X$ (i.e., T_n is strongly operator convergent with limit T). Then, the following inequality holds:*

$$\|T\| \leq \liminf_{n \to \infty} \|T_n\|.$$

For sequences in ℓ_1, we have the following result:

Theorem 2.4.3. *If $x_n \in \ell_1$ converges weak* to x and $\|x_n\|_1$ converges to $\|x\|_1$, then $\|x - x_n\|_1 \to 0$.*

2.4.1 Operators with Closed Range

In the following, some properties of bounded linear operators with closed range are stated.

Definition 2.4.4. *An operator $T \in \mathcal{B}(X, Z)$ has closed range if, for any $z \in Z$ for which there exists a sequence in the range of T converging to it, i.e.,*

$$\lim_{n \to \infty} \|z - z_n\| = 0, \qquad \text{with } z_n = Tx_n,\ x_n \in X,$$

there is a $x \in X$ such that $z = Tx$.

Theorem 2.4.4. *[22] If $T \in \mathcal{B}(X, Y)$, X and Y are Banach spaces and $\mathcal{R}(T)$ is closed, then*

$$\mathcal{R}(T^*) = [\mathcal{N}(T)]^{\perp}$$

Remark 2.4.1. If $T : X \to Y$ is onto, then, clearly, $\mathcal{R}(T) = Y$ is closed.

Since verifying that $\mathcal{R}(T)$ is closed in some cases can be difficult, it can be simpler to verify the equivalent conditions presented in the following result:

Theorem 2.4.5. *[23] If $T \in \mathcal{B}(X,Y)$ and X,Y are Banach spaces, then $\mathcal{R}(T)$ is closed in norm $\iff \mathcal{R}(T^*)$ is weak* closed $\iff \mathcal{R}(T^*)$ is closed in norm.*

2.5 Systems as Linear Operators on ℓ_∞^n

In this section, the systems are represented as operators between two signal spaces. In particular, we will consider linear time-invariant causal operators that map bounded sequences to bounded sequences.

Let P_t, $t \in Z^+$ be the truncation operator on $\ell_{\infty,e}^n$, i.e.,

$$P_t(x(0), x(1), \ldots) = (x(0), x(1), \ldots, x(t), 0, 0 \ldots)$$

Definition 2.5.1. *An operator T is said to be causal if*

$$P_t T = P_t T P_t \qquad \forall t$$

It is strictly causal if

$$P_t T = P_t T P_{t-1} \qquad \forall t$$

Let us denote by S the unit shift operator, i.e.,

$$S(x(0), x(1), \ldots) = (0, x(0), x(1), \ldots).$$

Definition 2.5.2. *The operator T is time-invariant if it commutes with the unit shift operator, i.e.,*

$$ST = TS.$$

Given a sequence $R(t) \in \mathbb{R}^{m\times n}$, R defines a linear, time-invariant, casual operator from ℓ_∞^n to ℓ_∞^m. This operator can be represented as a multiplication operator with an associated infinite matrix with a block lower triangular Toeplitz structure:

$$\begin{pmatrix} y(0) \\ y(1) \\ y(2) \\ \vdots \end{pmatrix} = \begin{pmatrix} R(1) & 0 & 0 & 0 & \ldots \\ R(1) & R(0) & 0 & 0 & \ldots \\ R(2) & R(1) & R(0) & 0 & \ldots \\ \vdots & \vdots & \vdots & \vdots & \ddots \end{pmatrix} \begin{pmatrix} x(0) \\ x(1) \\ x(2) \\ \vdots \end{pmatrix}, \tag{2.2}$$

where $x \in \ell_{\infty,e}^n$ and $y \in \ell_{\infty.e}^m$. Moreover, every linear, time-invariant, causal operator on $\ell_{\infty,e}^n$ can be represented in such a way. The space of these operators is denoted by $\mathcal{L}^{m\times n}$ and their action is simply indicated as $y = Rx$.

Notice that (2.2) gives a matrix representation of the convolution operator, where $R(t)$ is identified as the impulse response of the linear system:

$$y(t) = (R * x)(t) \triangleq \sum_{j=0}^{t} R(t-j)x(j).$$

2.5.1 Stability and Boundedness

Definition 2.5.3. *A linear system T is stable with respect to some input space X, and some output space Y, if it is bounded as a linear operator from X to Y.*

In other words, a system is stable if its maximum amplification of all possible inputs is finite.

Now, we specialize this definition to the cases of interest. Denote with $\mathcal{L}_{TI}{}^{m\times n}$ the space of all the bounded (stable) linear time-invariant (LTI) causal operators mapping ℓ_∞^n in ℓ_∞^m. Then, such a space can be characterized as follows:

Theorem 2.5.1. *$\mathcal{L}_{TI}{}^{m\times n}$ is given by the space of all block lower triangular matrices R such that*

$$\|R\| \triangleq \max_{1\le i\le m} \sum_{j=1}^{n} \sum_{t=0}^{\infty} |r_{ij}(t)| < \infty. \tag{2.3}$$

or, equivalently,

$$\|R\| = \max_{1\le i\le m} \sum_{j=1}^{n} \|r_{ij}\|_1.$$

which is nothing but the ℓ_1 norm of the impulse response sequence of the system. For this reason, the norm defined in Equation (2.3) is going to be denoted with $\|\cdot\|_1$.

Definition 2.5.4. *Given an element $R \in \mathcal{L}^{m\times n}$, its λ transform is defined in the following way:*

$$\hat{R}(\lambda) = \sum_{i=0}^{\infty} R(i)\lambda^i.$$

If $R \in \mathcal{L}_{TI}{}^{m\times n}$, then $\hat{R}(\lambda)$ is analytic in the open unit disc and continuous on the boundary. The collection of these elements, equipped with the norm defined in Equation (2.3), is traditionally denoted by $\mathbf{A}$ (o $\mathbf{A}^{m\times n}$). From this definition is clear that the spaces $\mathbf{A}^{m\times n}$, $\mathcal{L}_{TI}{}^{m\times n}$ and $\ell_1^{m\times n}$ are all isomorphic.

Definition 2.5.5. *Let $\mathcal{L}_\infty^{m\times n}$ denote the space of all complex-valued matrix functions on the unit circle that are bounded, i.e., if $\hat{R} \in \mathcal{L}_\infty^{m\times n}$, then*

$$\|\hat{R}\|_\infty = \text{ess}\ \sup_\theta \sigma_{max}\left[\hat{R}(e^{i\theta})\right] < \infty.$$

The subspace of $\mathcal{L}_\infty^{m\times n}$ of all the elements that admit analytic continuations in the unit disc is denoted by $\mathcal{H}_\infty^{m\times n}$, i.e.,if $\hat{R} \in \mathcal{H}_\infty^{m\times n}$, then $\hat{R}$ is analytic in the open unit disc and bounded on the unit circle.

The norm $\|\hat{R}\|_\infty$ for elements in $\mathcal{H}_\infty^{m\times n}$ is denoted by $\|\hat{R}\|_{\mathcal{H}_\infty}$. We can now characterize the bounded-energy input bounded-energy output stable LTI causal systems.

Theorem 2.5.2. *Every linear time-invariant causal bounded operator on $\ell_2(\mathbf{Z}_+)$ can be identified by a multiplication operator on $\mathcal{H}_2^n$ with an element $\hat{R} \in \mathcal{H}_\infty^{m\times n}$.*

2.5.2 Finite Dimensional LTI Systems

Definition 2.5.6. *A system $R \in \mathcal{L}^{p\times q}$ is said to be finite dimensional if it has a state space realization $(A, B, C, D) \in \mathbb{R}^{n\times n} \times \mathbb{R}^{n\times q} \times \mathbb{R}^{p\times n} \times \mathbb{R}^{p\times q}$, or, equivalently, if $\hat{R}(\lambda)$ is a real rational matrix.*

The state space realization of R is also denoted by

$$R = \left[\begin{array}{c|c} \mathrm{A} & \mathrm{B} \\ \hline \mathrm{C} & \mathrm{D} \end{array}\right].$$

Assuming that the realization in the state space is reachable and observable, then the system is ℓ_∞-stable (or ℓ_p-stable) if and only if all the eigenvalues of A are strictly inside the unit disc. The order of the system is given by the dimension of A. The Toeplitz matrix form relative to R is constructed by the impulse response:

$$R(t) = \begin{cases} D & \text{if } t = 0 \\ CA^{t-1}B & \text{if } t > 0 \end{cases}$$

Also the λ transform of R is easily calculated:

$$\hat{R}(\lambda) = \lambda C(I - \lambda A)^{-1}B + D$$

$\hat{R}^{\sim}(\lambda)$ denotes $\hat{R}^T(\frac{1}{\lambda})$ and is often called the Hilbert space adjoint of $\hat{R}$.

A more complete treatment of the subjects discussed in this section can be found in [25].

2.6 Rational Matrices

In this section, a few rational matrix theory concepts and results are presented. Here, LTI systems are seen as algebraic objects, represented by matrices of rational functions in λ (i.e. the λ transforms of the impulse response). Most of the results in this section can be found in [26] and [27].

Definition 2.6.1. *(**Unimodular Matrices**) A squared polynomial matrix $\hat{P}(\lambda) = P(0) + P(1)\lambda + \cdots + P(k)\lambda^k$ is said to be unimodular if its determinant is a non-zero constant independent of λ.*

Therefore, unimodular matrices have polynomial inverses and have full rank for all points in the complex plane. Moreover, the product of two unimodular matrices is a unimodular matrix.

Theorem 2.6.1. *Let $\hat{G}(\lambda)$ be a rational matrix $m \times n$ of normal rank (i.e., of rank r for almost all the λ). Then $\hat{G}(\lambda)$ can always be factorized in the following way:*

$$\hat{G}(\lambda) = \hat{L}(\lambda)\hat{M}(\lambda)\hat{R}(\lambda),$$

where $\hat{L}(\lambda)$ and $\hat{R}(\lambda)$ are unimodular matrices of appropriate dimensions, and

$$\hat{M}(\lambda) = \begin{pmatrix} \frac{\epsilon_1(\lambda)}{\psi_1(\lambda)} & & 0 & 0 & \cdots & 0 \\ & \ddots & & \vdots & \ddots & \vdots \\ & & \frac{\epsilon_r(\lambda)}{\psi_r(\lambda)} & 0 & \cdots & 0 \\ 0 & \cdots & 0 & 0 & \cdots & 0 \\ \vdots & \ddots & \vdots & \vdots & \ddots & \vdots \\ 0 & \cdots & 0 & 0 & \cdots & 0 \end{pmatrix}$$

is a matrix $m \times n$. The polynomials $\{\epsilon_i(\lambda), \psi_i(\lambda)\}$ are coprime for all the $i = 1, 2, \cdots, r$ and have the following property of divisibility: $\epsilon_i(\lambda)$ divides $\epsilon_{i+1}(\lambda)$ without remainder and $\psi_{i+1}(\lambda)$ divides $\psi_i(\lambda)$ without remainder, for $i = 1, 2, \cdots, r-1$.

$\hat{M}(\lambda)$ is a general canonical form for rational matrices known as the Smith-McMillan form. $\hat{M}(\lambda)$ is unique, while $\hat{L}(\lambda)$ and $\hat{R}(\lambda)$ are not.

2.6.1 Poles and Zeros

Definition 2.6.2. *Let $\hat{G}(\lambda)$ be a matrix of rational transfer functions with Smith-McMillan form $\hat{M}(\lambda)$. Then, the roots of $\prod_{i=1}^{r} \epsilon_i(\lambda)$ are the zeroes of $\hat{G}(\lambda)$ and the roots of $\prod_{i=1}^{r} \psi_i(\lambda)$ are the poles of $\hat{G}(\lambda)$.*

Definition 2.6.3. *Let λ_0 be a zero of $\hat{G}(\lambda)$. Let $\sigma_i(\lambda_0)$ be the multiplicity of λ_0 as a root of $\epsilon_i(\lambda)$. Then, $\{\sigma_i(\lambda_0)\}_{i=1}^{r}$ defines a non-decreasing sequence of non negative integers. For a given $i \in \{1, 2, \cdots, r\}$, $\sigma_i(\lambda_0)$ is known as the algebraic multiplicity of λ_0. The total number of indices i for which $\sigma_i(\lambda_0)$ is strictly positive is known as the geometric multiplicity of λ_0. Finally, $\sum_{i=1}^{r} \sigma_i(\lambda_0)$ defines the multiplicity of λ_0.*

Analogous definitions can be applied to the poles of $\hat{G}(\lambda)$, by allowing the σ_i's to assume non-positive integer values. The σ_i's are known as the structural indices of $\hat{G}(\lambda)$ at λ_0.

Corollary 2.6.1. *Let $\hat{G}(\lambda)$ be defined as in Theorem 2.6. Then, $\hat{G}(\lambda)$ has the following right and left polynomial factorizations:*

$$\hat{G}(\lambda) = \hat{D}_L^{-1}(\lambda)\hat{N}_L(\lambda) = \hat{N}_R(\lambda)\hat{D}_R^{-1}(\lambda),$$

where $\hat{N}_L(\lambda) = \hat{\mathcal{E}}(\lambda)\hat{R}(\lambda)$, $\hat{D}_L = \hat{\Psi}_L(\lambda)\hat{L}^{-1}(\lambda)$, $\hat{N}_R(\lambda) = \hat{L}(\lambda)\hat{\mathcal{E}}(\lambda)$, $\hat{D}_R(\lambda) = \hat{R}^{-1}(\lambda)\hat{\Psi}_R(\lambda)$ *and*

$$\begin{pmatrix} \epsilon_1(\lambda) & & 0 & 0 & \cdots & 0 \\ & \ddots & & \vdots & \ddots & \vdots \\ & & \epsilon_r(\lambda) & 0 & \cdots & 0 \\ 0 & \cdots & 0 & 0 & \cdots & 0 \\ \vdots & \ddots & \vdots & \vdots & \ddots & \vdots \\ 0 & \cdots & 0 & 0 & \cdots & 0 \end{pmatrix} \quad \textit{is } (m \times n)$$

$$\begin{pmatrix} \psi_1(\lambda) & & 0 & 0 & \cdots & 0 \\ & \ddots & & \vdots & \ddots & \vdots \\ & & \psi_r(\lambda) & 0 & \cdots & 0 \\ 0 & \cdots & 0 & 1 & \cdots & 0 \\ \vdots & \ddots & \vdots & \vdots & \ddots & \vdots \\ 0 & \cdots & 0 & 0 & \cdots & 1 \end{pmatrix} \quad \textit{is } (n \times n)$$

$$\begin{pmatrix} \psi_1(\lambda) & & 0 & 0 & \cdots & 0 \\ & \ddots & & \vdots & \ddots & \vdots \\ & & \psi_r(\lambda) & 0 & \cdots & 0 \\ 0 & \cdots & 0 & 1 & \cdots & 0 \\ \vdots & \ddots & \vdots & \vdots & \ddots & \vdots \\ 0 & \cdots & 0 & 0 & \cdots & 1 \end{pmatrix} \quad \textit{is } (m \times m)$$

Notice that the complete structure of the left (right) zeroes of $\hat{G}(\lambda)$ (i.e., the structural indices and the directions) is captured by $\hat{N}_R(\lambda)$ ($\hat{N}_L(\lambda)$). Moreover, every right (left) coprime polynomial factorization of $\hat{G}(\lambda)$ will have such property, while every right (left) coprime rational factorization will capture the structure of the left (right) unstable zeroes of $\hat{G}(\lambda)$.

2.7 Notation Summary

$\mathbf{Z}^+$ Set of the non-negative integer numbers.

$\mathbb{R}$ Set of the real numbers.

$\mathbb{R}^q$ Space of the $q-tuples$ of real numbers. If $x = (x_1, \ldots, x_q) \in \mathbb{R}^q$, then $|x|_\infty \triangleq \max_i |x_i|$.

$\mathbb{R}^{p\times q}$ Set of the real matrices with p rows and q columns.

ℓ_∞ Space of all the sequences of real numbers supported on the non negative integers. If $x \in \ell_\infty$, then $\|x\|_\infty \triangleq \sup_k |x(k)| < \infty$.

ℓ_∞^q Space of the $q-tuples$ of elements of ℓ_∞. If $x = (x_1, \cdots, x_q) \in \ell_\infty^q$, then $\|x\|_\infty = \max_i \|x_i\|_\infty$

$\ell_{\infty,e}^q$ Extended ℓ_∞^q space. It is equal to the space of all the $q-tuples$ of real numbers sequences.

ℓ_1 Space of the absolutely summable sequences supported over the non negative integers. If $x \in \ell_1$, then $\|x\|_1 = \sum_{k=0}^{\infty} |x(k)| < \infty$.

$\ell_1^{p\times q}$ Space of the $p \times q$ matrices with elements in ℓ_1. If $M = (m_{ij}) \in \ell_1^{p\times q}$, then $\|M\|_1 = \max_{1\le i\le p} \sum_{j=1}^{q} \|m_{ij}\|_1$.

P_k Truncation operator on sequences. If $x = \{x(i)\}_{i=0}^{\infty}$ is a sequence, then $P_k x = \{x(0), x(1), \ldots, x(k), 0, \ldots\}$.

S_k Right shift of k positions. If $x = \{x(i)\}_{i=0}^{\infty}$ is a sequence and k is a non negative integer, then $S_k x = \{\underbrace{0, \ldots, 0}_{k}, x(0), x(1), \ldots\}$.

3. Multi-Objective Control

This chapter contains the general control problem setup. It presents the affine characterization of all feasible closed loop maps and its alternative equivalent representation in terms of interpolation conditions. Several performance objectives and their representation as constraints on the closed loop system are also described. The material presented is well known and a more detailed presentation can be found in [1].

3.1 Control Problem Setup

We consider MIMO, discrete linear time-invariant (LTI) causal systems. Figure 3.1 shows a general set-up for posing performance specifications.

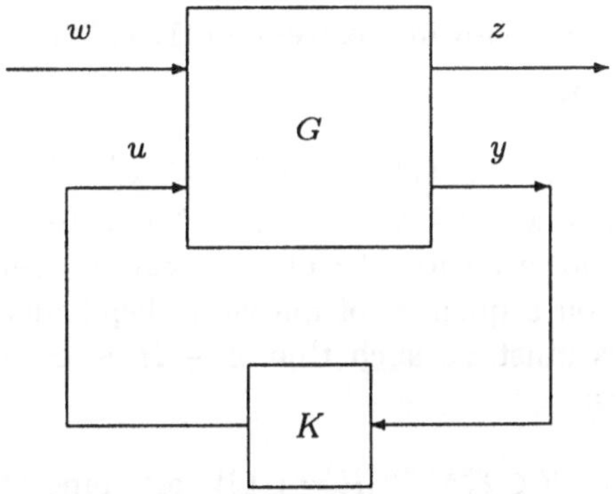

Fig. 3.1. General Set-Up

- u denotes the n_u *Control Inputs*,
- y denotes the n_y *Measured Outputs*,

- w denotes the n_w *Exogenous Inputs*,
- z denotes the n_z *Regulated Outputs.*

The standard augmented system is represented by the operator G, which is a 2×2 block matrix mapping the inputs w, u to the outputs z, y:

$$\begin{bmatrix} z \\ y \end{bmatrix} = \begin{bmatrix} G_{11} & G_{12} \\ G_{21} & G_{22} \end{bmatrix} \begin{bmatrix} w \\ u \end{bmatrix}.$$

The controller K uses the measured outputs and acts on the control inputs. We will assume that the closed loop system is well posed.

We are interested in the map between w and z, denoted by Φ:

$$\Phi = G_{11} + G_{12}K(I - G_{22}K)^{-1}G_{21}. \tag{3.1}$$

The general problem is to select a particular stabilizing controller for which the closed loop map Φ achieves the desired performance objectives.

Types of Problems.. When $n_u \geq n_z$ and $n_y \geq n_w$, we say that the problem is one-block, otherwise the problem is multi-block.

3.2 Feasibility Constraints

The set of all the BIBO stable closed loop maps Φ from w to z, each corresponding to an internally stable system with an LTI causal controller, is characterized by the following expression [1, 7, 28]:

$$\Phi = H - UQV, \tag{3.2}$$

where $H \in \ell_1^{n_z \times n_w}$, $U \in \ell_1^{n_z \times n_u}$ and $V \in \ell_1^{n_y \times n_w}$ are fixed stable systems depending on the plant, while $Q \in \ell_1^{n_y \times n_u}$ is a free parameter. The product UQV is a short notation for $U * Q * V$, where $*$ represents the convolution operation acting on sequences of matrices. Equivalently, the set of all stable closed loop maps must be such that $\Phi - H \in \mathcal{S}$, where the subspace $\mathcal{S}$ is defined as follows:

$$\mathcal{S} := \{R \in \ell_1^{n_z \times n_w} | R = UQV \text{ for some } Q \in \ell_1^{n_u \times n_y}\}.$$

The conditions on $R \in \ell_1^{n_z \times n_w}$, so that there exists a $Q \in \ell_1^{n_u \times n_y}$ with the property $R = UQV$, characterize the subspace $\mathcal{S}$, and are called interpolation conditions.

Assumption 3.2.1. *In the sequel it will be assumed, without loss of generality, that $\hat{U}(\lambda)$ has full column (normal) rank (i.e., rank of n_u for almost all λ) and $\hat{V}(\lambda)$ has full row rank (i.e, rank of n_y for almost all λ).*

Violation of this assumption implies that there are redundancies in the controls and/or in the measurements, which can be easily removed.

3.2.1 Interpolation Conditions

Consider the Smith-McMillan decomposition in Theorem (2.6) of the rational matrices $\hat{U}(\lambda)$ and $\hat{V}(\lambda)$,

$$\begin{aligned}\hat{U} &= \hat{L}_U \hat{M}_U \hat{R}_U \\ \hat{V} &= \hat{L}_V \hat{M}_V \hat{R}_V,\end{aligned} \tag{3.3}$$

where $\hat{L}_U$, $\hat{L}_V$, $\hat{R}_U$ and $\hat{R}_V$ are unimodular polynomial matrices and

$$\hat{M}_U = \begin{pmatrix} \frac{\epsilon_1}{\psi_1} & & \\ & \ddots & \\ & & \frac{\epsilon_{n_u}}{\psi_{n_u}} \\ 0 & \cdots & 0 \\ \vdots & \ddots & \vdots \\ 0 & \cdots & 0 \end{pmatrix}$$

$$\hat{M}_V = \begin{pmatrix} \frac{\epsilon'_1}{\psi'_1} & & & 0 \cdots 0 \\ & \ddots & & \vdots \ddots \vdots \\ & & \frac{\epsilon'_{n_y}}{\psi'_{n_y}} & 0 \cdots 0 \end{pmatrix}.$$

The structure of $\hat{M}_U$ and $\hat{M}_V$ reflects the assumption that $\hat{U}$ and $\hat{V}$ have full column and row rank respectively.

Let Λ_{UV} denote the set of the zeroes of $\hat{U}$ and $\hat{V}$ in $\overline{D}$ (the closed unit disc,). We make the following assumption, valid for the rest of this monograph:

Assumption 3.2.2. *$\hat{U}$ and $\hat{V}$ do not have zeroes on the unit circle, i.e. $\Lambda_{UV} \subset D$.*

Following the notation of Definition 2.6.3, let $\{\sigma_{U_i}(\lambda_0)\}_{i-1}^{n_y}$ denote the sequence of structural indices correspondent to $\hat{U}$ for every $\lambda_0 \in \Lambda_{UV}$. Similarly, define $\{\sigma_{U_i}(\lambda_0)\}_{i-1}^{n_y}$ for $\hat{V}$.

Consider the unimodular matrices $\hat{L}_U$ and $\hat{R}_V$. Since the inverse of these matrices are polynomial, one can define the following polynomial row and column vectors:

$$\begin{aligned}\hat{\alpha}_i(\lambda) &= (\hat{L}_U^{-1})_i(\lambda) \qquad & i &= 1, 2, \ldots, n_z \\ \hat{\beta}_j(\lambda) &= (\hat{R}_V^{-1})^j(\lambda) \qquad & j &= 1, 2, \ldots, n_w\end{aligned}$$

We are now ready to present the interpolation theorem [16].

Theorem 3.2.1. *Given $R \in \ell_1^{n_z \times n_w}$, there exists $Q \in \ell_1^{n_u \times n_y}$ such that $R = UQV$, if and only if, for every $\lambda_0 \in \Lambda_{UV} \subset D$, the following conditions are satisfied:*

$$i)\quad (\hat{\alpha}_i \hat{R} \hat{\beta}_j)^{(k)}(\lambda_0) = 0 \; for \; \begin{cases} i = 1, \ldots, n_u \\ j = 1, \ldots, n_y \\ k = 0, \ldots, \sigma_{U_i}(\lambda_0) + \sigma_{V_j}(\lambda_0) - 1 \end{cases} \tag{3.4}$$

$$ii)\quad \begin{cases} (\hat{\alpha}_i \hat{R})(\lambda) = 0 \; for \; i = n_u + 1, \ldots, n_z \\ (\hat{R} \hat{\beta}_j)(\lambda) = 0 \; for \; j = n_y + 1, \ldots, n_w \end{cases}$$

Theorem 3.2.1 establishes a set of algebraic conditions that are necessary and sufficient for R to be feasible (i.e. equivalent to UQV for some Q stable). The conditions in i), referred to as the *zero interpolation conditions*, assure that the structure of the unstable right and left zeroes of the composition UQV is maintained. The conditions in ii), referred to as the *rank interpolation conditions*, preserve the correct normal rank conditions. Such conditions are linear in R and can be easily rewritten in terms of linear functionals on R or Φ [16]. Thus, we have that $\Phi = H - UQV$, for some $Q \in \ell_1^{n_u \times n_y}$, if and only if Φ satisfies:

$$\begin{bmatrix} \mathcal{A}_Z \\ \mathcal{A}_R \end{bmatrix} \Phi = \begin{bmatrix} \mathcal{A}_Z \\ \mathcal{A}_R \end{bmatrix} H, \tag{3.5}$$

where $\mathcal{A}_Z : \ell_1^{n_z \times n_w} \to \mathbb{R}^{c_z}$ and $\mathcal{A}_R : \ell_1^{n_z \times n_w} \to \ell_1$ are the linear operators relative to the two conditions. $\mathcal{A}_Z$ has a finite dimensional range, since there is at most a finite number of zero interpolation constraints. $\mathcal{A}_R$ has an infinite dimensional range, because the rank interpolation conditions are represented by an infinite number of linear constraints on Φ.

By denoting $\mathcal{A}_{feas} = \begin{bmatrix} \mathcal{A}_Z \\ \mathcal{A}_R \end{bmatrix}$, and $b_{feas} = \begin{bmatrix} \mathcal{A}_Z \\ \mathcal{A}_R \end{bmatrix} H$, we obtain the following compact representation of Equation 3.5.

$$\mathcal{A}_{feas} \Phi = b_{feas}.$$

We must point out that there are other equivalent ways to describe the interpolation conditions, which result, however, in different operators $\mathcal{A}_{feas}$ [15]. When $\mathcal{A}_{feas}$ is constructed from the interpolation conditions of Theorem 3.2.1, the following result holds.

Theorem 3.2.2. *[15] $\mathcal{A}_{feas}$ has the property that its range is closed in $\mathbb{R}^{c_z} \times \ell_1$.*

Representation of the Interpolation Conditions. The interpolation conditions have more practical utility when viewed in the context of convolution of sequences, since this is the way they are going to be represented in the constrained optimization problems in this monograph. Following [16], for each zero interpolation condition, expanding the convolution operations we have that

$$(\hat{\alpha}_i \hat{R} \hat{\beta}_j)^{(k)}(\lambda_0) = \sum_{t=0}^{\infty} \left[\sum_{s=0}^{\infty} \sum_{l=0}^{\infty} \alpha_i(s-l) R(l) \beta_j(t-s) \right] (\lambda^t)^{(k)} \Bigg|_{\lambda=\lambda_0}$$
$$= \sum_{p=1}^{n_z} \sum_{q=1}^{n_w} \sum_{l=0}^{\infty} \left[\sum_{t=0}^{\infty} \sum_{s=0}^{\infty} \alpha_{iq}(s-l) \beta_{pj}(t-s) (\lambda^t)^{(k)} \right]_{\lambda=\lambda_0} r_{qp}(l).$$

For the rank interpolation conditions we have that

$$(\hat{\alpha}_i \hat{R})(\lambda) = 0 \Leftrightarrow (\alpha_i * R)(t) = \sum_{l=0}^{t} \alpha_i(t-l) R(l) = \underbrace{(0, \quad \cdots, \quad 0)}_{n_w} \quad \forall t \geq 0$$

$$(\hat{R} \hat{\beta}_j)(\lambda) = 0 \Leftrightarrow (R * \beta_j)(t) = \sum_{l=0}^{t} R(l) \beta_j(t-l) = \left. \begin{pmatrix} 0 \\ \vdots \\ 0 \end{pmatrix} \right\} n_z \quad \forall t \geq 0$$

Using these functional forms, we can obtain an equivalent representation of $\mathcal{A}_{feas}$ in terms of an infinite matrix, $A_{feas} : \ell_1 \to \ell_1$, by defining a reordering operator. The reordering operator, $\mathcal{O} : \ell_1^{n_z \times n_w} \to \ell_1$, takes the rows of $\Phi(k)$, for each k, and stacks them up into the infinite column vector ϕ, that is

$$\phi \triangleq \mathcal{O}\Phi \triangleq \begin{bmatrix} \phi_{11}(0) \\ \vdots \\ \phi_{1n_w}(0) \\ \phi_{21}(0) \\ \vdots \\ \phi_{2n_w}(0) \\ \vdots \\ \phi_{n_z 1}(0) \\ \vdots \\ \phi_{n_z n_w}(0) \\ \phi_{11}(1) \\ \vdots \end{bmatrix} \in \ell_1$$

The operator $\mathcal{O}$ is linear, one-to-one, onto, and its inverse is equal to its adjoint, $\Phi = \mathcal{O}^* \phi$. The matrix representation of $\mathcal{A}_{feas}$ is given by: $A_{feas} = \mathcal{A}_{feas} \mathcal{O}^*$. The matrix representation of $\mathcal{A}_{feas}$ allows us to write down each one of the feasibility constraints on Φ in terms of scalar product of the rows of A_{feas} with ϕ.

Remark 3.2.1. Notice that a consequence of the Assumption 3.2.2 is that the rows of $A_{feas} : \ell_1 \to \ell_1$ are elements of c_0 [15].

Comment: In this section we have seen that the achievable closed loop maps can be represented either through the Youla parametrization, $\Phi = H - UQV$, or through the interpolation conditions, $\mathcal{A}_{feas}\Phi = b_{feas}$. Both representations are in terms of infinite linear equality constraints, although, the interpolation conditions reduce to a finite number of constraints in one-block problems. Mathematically, however, they provide different points of view of the same control problem, and they lead to different computational methods with different properties. Therefore, when approaching a new problem, it is advisable to carry out the derivations with both representations and to choose the one with the more desirable properties. Although in this monograph, the interpolation conditions point of view is dominant, a good example of the use of the properties of both representations is given in Chapter 8.

3.3 Control Objectives

Nominal stability of the closed loop system is only one of the many objectives that the closed loop must satisfy.

3.3.1 Performance with Fixed Inputs

In many control applications, it is desired to track a specific trajectory with the least overshoot, or undershoot, possible. In a more general setting, the controller is designed so that the response to a fixed input is within a time-domain template:

$$g(t) \leq z(t) \leq h(t)$$

with both g, h specified. We will say z is feasible if $z = \Phi * w_0$ and Φ is feasible. Constraints on the overshoot, undershoot, settling time, and maximum deviation are captured by this representation.

All of the specifications above result in linear constraints on the closed loop map. By combining all of these constraints, a linear operator $\mathcal{A}_{temp}$ can be constructed such that the exact input specifications are equivalent to:

$$\mathcal{A}_{temp}\Phi \leq b_{temp}$$

for some fixed b_{temp}. Notice that the range of $\mathcal{A}_{temp}$ may be infinite dimensional. For example, let the specification be that the response to the step input is always smaller then $\gamma \geq 0$. Then, in the SISO case, $\mathcal{A}_{temp}$ and b_{temp} are given by the following infinite arrays

$$\mathcal{A}_{temp} = \begin{bmatrix} 1 & 0 & 0 & \cdots & 0 & \cdots \\ 1 & 1 & 0 & \cdots & 0 & \cdots \\ 1 & 1 & 1 & \ddots & \ddots & \ddots \\ \vdots & \ddots & \ddots & \ddots & & \end{bmatrix} \qquad b_{temp} = \gamma \cdot \begin{bmatrix} 1 \\ 1 \\ 1 \\ \vdots \end{bmatrix}$$

3.3.2 Average and Worst-Case Performance

Some performance objectives are naturally expressed in terms of constraints on some norm of the closed loop system. For example, worst-case peak-to-peak gain is captured by the ℓ_1 norm of the closed loop system. Performance on the response to white-noise input is usually expressed in terms of variance of the output signal, and is captured by constraining the ℓ_2 norm of the closed loop. These constraints are convex on Φ and are expressed as

$$\|\Phi\|_1 \leq \gamma_1, \qquad \|\Phi\|_2 \leq \gamma_2.$$

In general, these constraints can act on separate elements of the closed loop transfer function matrix.

3.3.3 Performance in the Frequency Domain

Often, performance objectives are given as desired characteristics of the closed loop frequency response. Constraints on the magnitude of the frequency response are captured by the weighted $\mathcal{H}_\infty$ norm of the closed loop:

$$\|\hat{W}\hat{\Phi}\|_{\mathcal{H}_\infty} \leq \gamma,$$

where according to Definition 2.5.5

$$\|\hat{W}\hat{\Phi}\|_{\mathcal{H}_\infty} = \text{ess} \sup_{\omega \in [0,2\pi)} \sigma_{max} \left[\hat{W}(e^{j\omega})\hat{\Phi}(e^{j\omega})\right].$$

In some cases, adjustments only in some localized frequency regions are needed to achieve the desired shape. In such situations, magnitude constraints at some frequency points can be directly imposed on the closed loop to perform the desired shaping. Such constraints are given as follows:

$$|\hat{\Phi}(e^{j\omega_i})| \leq \gamma_i, \quad i = 1, \ldots, n \tag{3.6}$$

The $\mathcal{H}_\infty$ constraints, as well as the magnitude constraints at discrete frequency points, are convex constraints on Φ. More precisely, they are LMI constraints on Φ.

3.3.4 LMI Constraints

Definition 3.3.1. *Given an element $P \in \mathbb{R}_S^{m\times m}$, P is positive semi-definite if $v^T P v \geq 0$, for all $v \in \mathbb{R}^m$. When $P \in \mathbb{R}_S^{\infty}$ and belongs to an LMI space, P is positive-definite if $v^T P v \geq 0$ for all $v \in \ell_2$.*

Note that the positive semi-definite matrices of an LMI space form a cone according to Definition 2.2.2. It is with respect to this cone that the matrix inequalities are defined. Given S and R in a LMI space, $S \geq R$ if $S - R$ is positive semi-definite.

A Linear Matrix Inequality is given by the following relation

$$S(x) \geq S^o,$$

where $S(x)$ and S^o are elements of an LMI space (symmetric matrices), $S(x)$ depends linearly on some variable x belonging to a vector space, and the inequality is with respect to the cone of positive semi-definite symmetric matrices.

Examples. An example of a finite-size infinite dimensional LMI is given by a magnitude constraint on the closed loop transfer function at a frequency point ω_0. The magnitude constraint in (3.6) for $\Phi \in \ell_1$ and $n = 1$ is equivalent to the following LMI constraint:

$$\begin{bmatrix} \beta I & \hat{\Phi}(e^{j\omega_0}) \\ \hat{\Phi}^*(e^{j\omega_0}) & \beta I \end{bmatrix} \geq 0$$

LMI's on complex Hermitian matrices can be rearranged to LMI's on real symmetric matrices as follows.

$$\mathcal{S}(\Phi) + S^C \geq 0 \Leftrightarrow \sum_{i=1}^{\infty} S(i)\Phi(i) + S^C \geq 0 \tag{3.7}$$

where

$$S(i) = \begin{bmatrix} 0 & \cos(i\omega_0) & 0 & -\sin(i\omega_0) \\ \cos(i\omega_0) & 0 & \sin(i\omega_0) & 0 \\ 0 & \sin(i\omega_0) & 0 & \cos(i\omega_0) \\ -\sin(i\omega_0) & 0 & \cos(i\omega_0) & 0 \end{bmatrix} \tag{3.8}$$

and

$$S^C = \beta \begin{bmatrix} 1 & 0 & 0 & 0 \\ 0 & 1 & 0 & 0 \\ 0 & 0 & 1 & 0 \\ 0 & 0 & 0 & 1 \end{bmatrix}. \tag{3.9}$$

An infinite size infinite dimensional LMI constraint is given by a constraint on the ℓ_2 norm of the closed loop map $\Phi \in \ell_1$:

$$\|\Phi\|_2 \leq \beta,$$

which is equivalent to

$$\begin{bmatrix} \beta & \Phi^T \\ \Phi & \beta I \end{bmatrix} \geq 0.$$

Convex optimization problems with linear cost and LMI constraints are called LMI problems. LMI problems subject to finite dimensional LMI constraints (X is finite dimensional and $\mathcal{S}(x) \in \mathbb{R}_S^{m \times m}$) can be solved efficiently with interior point methods, and are common in many problems in control theory and in engineering. For a detailed description of finite dimensional LMI problems in control theory see [8].

Much less is known about the solvability of infinite dimensional, possibly infinite-size, LMI problems. The approach presented in this monograph can be used to analyze such problems, and derive computable approximations.

Summary and Comments. The stability constraints and the constraints from several performance objectives can all be included into an optimization problem. Many different design problems can be posed in this way. In the next chapter we describe how all these possible optimization problems can be analyzed in a unified way.

4. Generalized Linear Programs and Duality Theory

In this chapter, we present a unified treatment for most convex optimization problems that arise from multi-objective control problems. In the previous chapter, we have presented several performance objectives of practical importance and described how they translate into sets of convex constraints on the closed loop map. In this chapter we will see that, the common characteristic of these sets is that each one can be seen, in an extended space, as a cone possibly intersected with a translated subspace. Also the set of feasible set, intersection of all constrained sets, can be seen as a cone in an extended space, possibly intersected by a translated subspace. This property allows us to rewrite the multi-objective control problem as a generalized linear program subject to linear equality constraints and with the variables belonging to a cone. Thus, duality needs to be studied only for this class of problems. The primal-dual pair of convex optimization problems is transformed into a primal-dual pair of linear programming problems and the duality relationship is preserved. Although duality theory for infinite dimensional linear programs is well developed, in this chapter, we follow [29] and present some useful results that exploit certain common features of the optimization problems of interest.

4.1 Control Problems as Linear Programs

It is interesting to note that many convex optimization problems can be represented as standard linear programming problems, with equality constraints, and an appropriate positive cone in the primal space. To see this, we first analyze the kind of constraints most commonly encountered. For clarity of exposition, we present simplified examples. The same ideas apply to more general cases.

Pointwise linear inequalities:

Let $\mathcal{A} \in \mathcal{B}(\ell_1, \ell_\infty)$, then

$$\mathcal{A}x \leq b \Leftrightarrow \begin{cases} \mathcal{A}x + z = b \\ (x,z) \in P \\ P = \{(x,z) \in \ell_1 \times \ell_\infty \,|\, z_i \in \mathbb{R}^+ ,\, i = 1,2,\ldots\} \end{cases}$$

Note that P is in an extended space, and that the constraints are given by the intersection of P with the linear variety

$$\{(x, z) \in \ell_1 \times \ell_\infty \mid Ax + z = b\}$$

Norm constraints:

$$\|x\|_p \le \gamma \Leftrightarrow \begin{cases} \xi = \gamma \\ (x, \xi) \in -P \\ P = \left\{(x, \xi) \in \ell_p \times \mathbb{R} \mid \|x\|_p - \xi \ge 0,\ \xi \ge 0\right\} \end{cases}$$

Also in this case, the constraint is represented as the intersection of a cone and a linear variety in an extended space.

LMI constraints:

Let $S(x) : X \to \mathbb{R}_S^{m \times m}$ be linear in x, and $S^o \in \mathbb{R}_S^{m \times m}$.

$$S(x) \le S^o \Leftrightarrow \begin{cases} S(x) + R = S^o \\ (x, R) \in P \\ P = \left\{(x, R) \in X \times \mathbb{R}_S^{m \times m} \mid R \text{ is positive semi-definite } \right\} \end{cases}$$

This representation can be generalized to infinite-size LMI.

4.1.1 Example

We now present a simple example to show the mechanics of this transformation. Consider the following mixed objective optimization:

$$\begin{aligned} \mu^0 = \quad & \inf_{x \in \ell_1} \quad \|x\|_1 \\ & Ax = b \\ & \|x\|_2 \le c \end{aligned}$$

From the previous development, the constraint $\|x\|_2 \le c$ is equivalent to $(x, \xi_1, \xi_2) \in P_2$, with $\xi_2 = c$, where

$$P_2 = \{(x,\ \xi_1,\ \xi_2) \in \ell_1 \times \mathbb{R} \times \mathbb{R} \mid -\|x\|_2 + \xi_2 \ge 0\}.$$

Analogously, the constraint $\|x\|_1 \le \xi$ is expressed as $(x, \xi_1, \xi_2) \in P_1$, where

$$P_1 = \{(x,\ \xi_1,\ \xi_2) \in \ell_1 \times \mathbb{R} \times \mathbb{R} \mid -\|x\|_1 + \xi_1 \ge 0\}.$$

Clearly, the variables must belong to the intersection of the set $P = P_1 \cap P_2$ and $\{(x, \xi_1 \xi_2) \in \ell_1 \times \mathbb{R} \times \mathbb{R} \mid Ax = b\}$, for the problem to be feasible.

Then, the equivalent linear program is given by

$$\begin{aligned} \mu^0 = \quad & \inf_{x \in \ell_1, \xi_1, \xi_2 \in \mathbf{R}} \quad \xi_1 \\ & Ax = b \\ & \xi_2 = c \\ & (x,\ \xi_1,\ \xi_2) \ge 0\ (\in P) \end{aligned}$$

Several other transformations will be presented in the next chapters.

4.2 Duality Theory Results

The problems we focus on can be represented as generalized linear programs:

$$\begin{aligned} &\inf_{x \in X} \quad \langle x, c \rangle \\ &\mathcal{A}x = b \\ &x \leq 0 \end{aligned} \tag{4.1}$$

where $\mathcal{A} \in \mathcal{B}(X, Z)$ and the inequality $x \leq 0$ is defined with respect to some positive cone P. The Lagrangian duality theory establishes a duality correspondence between the primal problem in the space X and a dual problem in the dual of the constraint space. The constraint space is the space where the image of the constraint operator lies. The main result of the theory of Lagrange multipliers for convex optimization problems is given in the next theorem.

The introduction of positive cones allows to generalize the definition of convexity as well.

Definition 4.2.1. *Given the positive cone P in Z, a mapping $F : X \to Z$ is convex if the domain Ω of F is convex and if $F(\alpha x_1 + (1-\alpha)x_2) \leq \alpha F(x_1) + (1-\alpha)F(x_2)$, for all x_1, $x_2 \in \Omega$ and all $\alpha \in (0,1)$, where the inequality is defined with respect to the positive cone P.*

Theorem 4.2.1. *[22] Let X be a linear vector space, Z a normed space, Ω a convex subset of X and P the positive cone in Z. Assume that P contains an interior point. Let f be a real-valued convex functional on Ω and G a convex mapping from Ω into Z. Assume the existence of a point $x_1 \in \Omega$ for which $G(x_1) < 0$ (i.e., $G(x_1)$ is an interior point of $N = -P$).*
Let

$$\mu^o = \inf_{\substack{x \in \Omega \\ G(x) \leq 0}} f(x) \tag{4.2}$$

and assume that μ^o is finite. Then,

$$\mu^o = \max_{z^* \geq 0} \inf_{x \in \Omega} [f(x) + \langle G(x), z^* \rangle],$$

and the maximum on the right is achieved by some $z_0^* \geq 0 \in Z^*$ (where the inequality is defined with respect to the positive cone $P^{\oplus}$). If the infimum in (4.2) is achieved by some $x_0 \in \Omega$, then,

$$\langle G(x_0), z_0^* \rangle = 0$$

and x_0 minimizes $f(x) + \langle G(x), z_0^* \rangle$, with $x \in \Omega$.

Although an equality constraint $H(x) = 0$ is equivalent to two inequality constraints $G(x) = \begin{bmatrix} H(x) \\ -H(x) \end{bmatrix} \leq 0$, there is no $x_1 \in \Omega$ such that $G(x_1) < 0$. Therefore, one of the hypotheses of the theorem is never satisfied. Nevertheless, the above theorem can still be applied to the problem in (4.1) by including the equality constraints $\mathcal{A}x - b = 0$ directly in the definition of the set Ω i.e., by setting $\Omega = \{x \in X \mid \mathcal{A}x = b\}$. Then, the following result holds:

Theorem 4.2.2. *Let X, Z be Banach spaces and P be the positive cone in X. Assume that there exists an $x_1 \in X$ such that $\mathcal{A}x_1 = b$ and $x_1 < 0$, i.e., $-x_1$ is an interior point of P. Assume, moreover, that $\mathcal{A} \in \mathcal{B}(X, Z)$ has closed range and that the optimal value μ^o for Problem (4.1) is finite. Then,*

$$\mu^o = \inf_{\substack{x \in X \\ \mathcal{A}x = b \\ x \leq 0}} \langle x, c \rangle = \max_{\substack{z^* \in Z^*, \lambda \in X^* \\ -\mathcal{A}^* z^* + c + \lambda = 0 \\ \lambda \geq 0}} \langle b, z^* \rangle$$

where the inequality $\lambda \geq 0$ is defined with respect to the cone $P^\oplus$, the maximum is achieved for some $z_o^* \in Z^*$ and $\lambda_0 \in P^\oplus \subset X^*$. If the infimum on the left is achieved by some $x_o \in X$, then,

$$\langle x_o, c \rangle = \langle x_o, \mathcal{A}^* z_o^* \rangle.$$

Proof. By applying Theorem 4.2.1 with $\Omega = \{x \in X \mid \mathcal{A}x = b\}$ and $G(x) = x$, we have that:

$$\mu^o = \max_{\lambda \geq 0} \inf_{\mathcal{A}x = b} (\langle x, c \rangle + \langle x, \lambda \rangle)$$

This problem has a finite value only if $c + \lambda$ annihilates all the elements in $\mathcal{N}(\mathcal{A})$. Since $\mathcal{R}(\mathcal{A})$ is closed, $\mathcal{N}(\mathcal{A})^\perp = \mathcal{R}(\mathcal{A}^*)$ (Thm. lunaperp.thm). Thus, we have that

$$c + \lambda = \mathcal{A}^* z^*, \qquad \text{for } z^* \in Z^*$$

and the result follows by using the properties of the adjoint operator. ∎

Comment: Notice that the well known duality result of Theorem 3.12 in [30], which deals exactly with Problem (4.1) is a special case of the result of Theorem 4.2.2. Theorem 3.12 in [30] requires $\mathcal{A}$ to be onto Z and weakly-continuous. This implies, respectively, that $\mathcal{A}$ is bounded ([23] Thm 1.1 pg 171), and it has closed range in Z. Thus, by selecting Z to be the range of $\mathcal{A}$, we can recover the result of Theorem 3.12 in [30]. Moreover, the adjoint of $\mathcal{A}$ from Z^* to X^* may have a nicer representation than the adjoint of $\mathcal{A}$ from $(\mathcal{R}(\mathcal{A}))^*$ to X^*.

In some instances, the minimization problem is already the dual of some maximization problem. In this case the minimizing solution is guaranteed to exist, while the maximizing solution may not. Next corollary refers to this situation.

Corollary 4.2.1. *Let X be the dual of some Banach space Y, and $\mathcal{A} \in \mathcal{B}(Z, Y)$. Assume that there is a $z_1 \in Z$ such that $\mathcal{A}z_1 - c > 0$, i.e., $\mathcal{A}z_1 - c$ belongs to the interior of the positive cone defining the relation $\mathcal{A}z - c \geq 0$. Then, if μ^o is finite,*

$$\min_{\substack{\mathcal{A}^* x = b^* \\ x \leq 0}} \langle c, x \rangle = \sup_{\mathcal{A}z - c \geq 0} \langle z, b^* \rangle = \mu^o \tag{4.3}$$

where $x \leq 0$ if and only if $-x \in P^{\oplus}$ (or $x \in P^{\ominus}$).

Proof. Add a slack variable $y \geq 0 \in Y$. Reformulate the maximization problem on the right-hand side of the above equation in $Z \times Y$ as follows:

$$\sup_{\mathcal{A}z - c \geq 0} \langle z, b^* \rangle = -\inf_{\substack{[-\mathcal{A}\ I] \begin{bmatrix} z \\ y \end{bmatrix} = -c \\ z \in Z, y \geq 0}} \langle z, -b^* \rangle + \langle y, 0 \rangle$$

where the positive cone in $Z \times Y$ is now given by $Z \times P$ and it has a nonempty interior, since P does. Note that the operator $[-\mathcal{A}\ I]$ automatically has a closed range in Y. The result then follows from Theorem 4.2.2, with $x \geq 0$ instead of $x \leq 0$ and simple rearrangements. ∎

The result of Theorem 4.2.2 may not be very useful in general, since it requires the computation of $P^{\oplus}$. In many cases, the positive cone P is given by the intersection of other simpler cones, say $P = \cap P_i$, resulting from various performance specifications. It is well known ([31]) that $P^{\oplus} = (\cap P_i)^{\oplus} = cl \sum P_i^{\oplus}$, where *cl* indicates the *weak** closure. However, the main result of this chapter, presented next, shows that, in these cases, it is sufficient to consider the set $\sum P_i^{\oplus}$ as the positive cone in the dual space.

Theorem 4.2.3. *Let X, Z be Banach spaces and P be a cone in X given by the intersection of a finite number of cones P_i in X, $P = \cap_{i=1}^n P_i$. Assume that there exists an $x_1 \in X$ such that $\mathcal{A}x_1 = b$ and $-x_1$ is an interior point of P. Assume that $\mathcal{A} \in \mathcal{B}(X, Z)$ has closed range and that the optimal value μ^o for Problem (4.1) is finite. Then,*

$$\mu^o = \inf_{\substack{x \in X \\ \mathcal{A}x = b \\ -x \in P}} \langle x, c \rangle = \max_{\substack{z^* \in Z^*, \lambda \in X^* \\ -\mathcal{A}^* z^* + c + \lambda = 0 \\ \lambda \in \sum_{i=1}^n P_i^{\oplus}}} \langle b, z^* \rangle$$

and the maximum is achieved for some $z_o^* \in Z^*$ and $\lambda_0 \in \sum_{i=1}^n P_i^{\oplus} \subset X^*$. If the infimum on the left is achieved by some $x_o \in X$, then,

$$\langle x_o, c \rangle = \langle x_o, \mathcal{A}^* z_o^* \rangle$$

Proof. Consider the set $\Omega_1 = \{x \in X \mid \mathcal{A}x = b,\ -x \in P_i,\ i = 1, \ldots, n-1\}$ and let P_n define the positive cone. Then, from Theorem 4.2.1, we have that

$$\mu^o = \inf_{\substack{x \in \Omega_1 \\ -x \in P_n}} \langle x, c \rangle = \max_{\lambda_n \in P_n^{\oplus}} \inf_{x \in \Omega_1} \langle x, c + \lambda_n \rangle$$

Next, define $\Omega_2 = \{x \in X \mid \mathcal{A}x = b,\ -x \in P_i,\ i = 1, \ldots, n-2\}$ and let the positive cone be P_{n-1}. Apply the theorem again to

$$\inf_{x\in\Omega_1}\langle x, c+\lambda_n\rangle = \inf_{x\in\Omega_2,\, -x\in P_{n-1}}\langle x, c+\lambda_n\rangle$$

we get,

$$\mu^o = \max_{\lambda_n\in P_n^{\oplus}}\ \max_{\lambda_{n-1}\in P_{n-1}^{\oplus}}\ \inf_{x\in\Omega_2}\langle x, c+\lambda_n+\lambda_{n-1}\rangle = \max_{\lambda\in P_n^{\oplus}+P_{n-1}^{\oplus}}\ \inf_{x\in\Omega_2}\langle x, c+\lambda\rangle$$

Repeat the procedure up to $n-1$ and apply the result in Theorem 4.2.2 at the last step. The result then follows. ■

4.2.1 Example: Mixed ℓ_1/Finite-size LMI Problem

Consider the following optimization problem

$$\begin{aligned} \mu^o = \quad & \inf \quad \|\Phi\|_1 \\ \text{subject to:} \quad & \\ & \mathcal{A}\Phi = b \\ & \mathcal{S}(\Phi) \geq S^C \end{aligned}$$

where $\mathcal{S}(\Phi) \geq S^C$ is a finite-size LMI constraint, i.e., $\mathcal{S} \in \mathcal{B}(\ell_1, \mathbb{R}_S^{m\times m})$ with '$\geq$' meaning with respect to the cone of positive definite symmetric matrices, and $\mathcal{A}\Phi = b$ is a set of linear constraints , with $\mathcal{A} \in \mathcal{B}(\ell_1, \mathbb{R}^n)$, representing the interpolation conditions. A Special case of this problem is discussed in Chapter 6.

We remove the inequality by introducing a slack variable $H \in \mathbb{R}_S^{m\times m}$. The problem can be rewritten as follows:

$$\begin{aligned} \mu^o = \quad & \inf \quad \gamma \\ \text{subject to:} \quad & \\ & \mathcal{A}\Phi = b \\ & \mathcal{S}(\Phi) + H = S^c \\ & -(\Phi, \gamma, H) \in P \end{aligned}$$

where $P = P_1 \cap P_2$, with

$$P_1 = \{(x, y, z) \in \ell_1 \times \mathbb{R} \times \mathbb{R}_S^{m\times m} \mid -\|x\|_1 - y \geq 0, -y \geq 0\}$$

and

$$P_2 = \{(x, y, z) \in \ell_1 \times \mathbb{R} \times \mathbb{R}_S^{m\times m} \mid z \geq 0\}$$

The conjugate cones are given by the following sets

$$P_1^{\oplus} = \{(\Phi_1^*, \gamma_1^*, H_1^*) \in \ell_\infty \times \mathbb{R} \times \mathbb{R}_S^{m\times m} \mid \|\Phi_1^*\|_\infty \leq -\gamma_1^*,\ \gamma_1^* \leq 0,\ H_1^* = 0\}$$

$$P_2^{\oplus} = \{(\Phi_2^*, \gamma_2^*, H_2^*) \in \ell_\infty \times \mathbb{R} \times \mathbb{R}_S^{m\times m} \mid \Phi_2^* = 0,\ \gamma_2^* = 0,\ H_2^* \geq 0\}$$

The operator representing all the equality constraints is onto $\mathbb{R}^n \times \mathbb{R}_S^{m \times m}$, and therefore it has closed range. Now, if the regularity conditions are satisfied, then the dual problem with no duality gap is given by the following optimization

$$
\begin{aligned}
\mu^o = \quad & \max \quad \langle z_1^*, b \rangle + Trace(H^* S^{C}) \\
\text{subject to:} \quad & \\
& \|\mathcal{A}^* z_1^* + \mathcal{S}^*(H^*)\|_\infty \leq 1 \\
& H^* \geq 0 \\
& z_1^* \in \mathbb{R}^n, \; H^* \in \mathbb{R}_S^{m \times m}
\end{aligned}
$$

where $\mathcal{S}^*(H^*)$ is a sequence in ℓ_∞ with each component

$$(\mathcal{S}^*(H^*))_n = Trace(H^* S(n))$$

To derive the dual problem we have substituted the expressions for the conjugate cones in Theorem 4.2.3, rearranged, used the fact that the adjoints of $\mathcal{A}$ and $\mathcal{S}$ are equal to their transposed, and used the representation of linear functionals on $\mathbb{R}_S^{m \times m}$ given in Remark 2.1.1.

Summary and Comments. The approach to study the duality relationship for the multi-objective optimization problems can be summarized as follows. The given convex optimization problem is first transformed into a linear programming problem. Then, duality theory is applied to the linear program to construct the dual program and study its properties. Finally, the dual convex problem is obtained, by transforming back the abstract dual linear program into the equivalent convex optimization problem. In the next chapters we apply this general approach and the results of this chapter to the study of several multi-objective control problems of practical importance.

5. ℓ_1 Optimal Control with Constraints on the Step Response

This is the first of three chapters which contain applications of the theory developed so far to important multi-objective problems. The treatment presented shows that the approach proposed is successful in analyzing the solvability of these problems. However, the application of the results of the previous chapter to the given problem may not be so immediate. The different issues involved and the possible difficulties in the application of the theory are highlighted throughout.

In this chapter, we solve the following problem: for a given MIMO plant, find a stabilizing controller which minimizes the ℓ_1 norm of the closed loop map and such that the response z to some fixed input $w \in \ell_\infty^{n_w}$ is constrained to lie in some time domain template. We study the case where the extra constraints are imposed over a finite-time horizon, and the extension to infinite horizon. Several results are derived and the treatment is rather complete. We present a design example and show the utility of the computational approach in deriving fundamental limitations of the closed loop system and tradeoffs in the achievable performance.

5.1 Problem Definition

In many control applications, one is interested in tracking a specific trajectory with the least overshoot (or undershoot) possible, while still maintaining good disturbance rejection and robustness properties. Thus, in addition to minimizing the ℓ_1-norm of some closed loop map, the controller must be designed so that the response to a fixed input w is within a time-domain template:

$$g(t) \leq z(t) \leq h(t)$$

with both g and h specified and $z = \Phi * w$. These constraints are linear on Φ ([1]). The operator $\mathcal{A}_w$, representing the convolution with w, has a matrix form with a block Toeplitz structure. The output z is given by $z = \mathcal{A}_w \Phi = \Phi * w$. Note that, for each finite i, the ith row of A_w (the matrix representation of $\mathcal{A}_w$), is an element in c_0.
For example, in the SISO case and for w equal to a step sequence, $\mathcal{A}_w$ is given by:

$$\mathcal{A}_w = \begin{bmatrix} 1 & 0 & 0 & \cdots & 0 & \cdots \\ 1 & 1 & 0 & \cdots & 0 & \cdots \\ 1 & 1 & 1 & \ddots & \ddots & \ddots \\ \vdots & \ddots & \ddots & \ddots & & \end{bmatrix}$$

In a general setting, we may have to constrain the responses to several fixed inputs. It is possible to construct a linear operator $\mathcal{A}_{temp}$ such that the exact output specifications are equivalent to:

$$\mathcal{A}_{temp}\Phi \leq b_{temp}$$

for some fixed b_{temp}.
For the sake of simplicity, and without loss of generality, we consider the case $\mathcal{A}_{temp} = \mathcal{A}_w$.
If the constraints are imposed only up to a finite time (finite-horizon time-domain template), only a finite number of constraints is needed and the operator $\mathcal{A}_{temp}$ maps $\ell_1^{n_z \times n_w}$ to $\mathbb{R}^n$ for some n.

5.2 The Finite-Horizon Case

We assume that we need to apply the constraints up to a finite time T. This problem is the simplest generalization of the standard ℓ_1 problem and it retains all the properties of the standard ℓ_1 optimization.

Theorem 5.2.1. *Consider the problem:*

$$\mu^o = \inf_{\substack{\mathcal{A}_{feas}\Phi = b_{feas} \\ \mathcal{A}_{temp}\Phi \leq b_{temp}}} \|\Phi\|_1 \tag{5.1}$$

where $\Phi \in \ell_1^{n_z \times n_w}$, $\mathcal{A}_{feas} : \ell_1^{n_z \times n_w} \to \ell_1$ defined as in Equation (3.5), and $\mathcal{A}_{temp} : \ell_1^{n_z \times n_w} \to \mathbb{R}^n$ represents the convolution with some fixed input $w \in \ell_\infty^{n_w}$.

The inequalities are defined with respect to the natural positive cone in $\mathbb{R}^n$. Assume further that the problem is feasible and stays feasible if the inequality constraints are slightly tightened, i.e., there is an $\epsilon > 0$ such that the problem is still feasible when the vector b_{temp} is replaced by $b_{temp} - 1\epsilon$, where 1ϵ is a column vector of suitable dimension with each component equal to ϵ. Then, there is no duality gap between Problem (5.1) and its Lagrangian dual.

Proof. To see how Problem (5.1) can be transformed in the form of problem (4.1), define the following cones in $X = \ell_1^{n_z \times n_w} \times \mathbb{R} \times \mathbb{R}^n$:

$$N_1 = \{\Phi, \xi_1, \xi_2, \mid \|\Phi\|_1 - \xi_1 \leq 0,\ \xi_1 \geq 0\}$$

and

$$N_2 = \{\Phi, \xi_1, \xi_2 \mid \xi_2 \geq 0\}$$

. Let $P = -N_1 \cap -N_2$ be the positive cone.

Then, Problem (5.1) is equivalent to:

$$\mu^o = \inf \xi_1$$
$$\text{subject to: } \begin{bmatrix} \mathcal{A}_{feas} & 0 & 0 \\ \mathcal{A}_{temp} & 0 & I \end{bmatrix} \begin{bmatrix} \Phi \\ \xi_1 \\ \xi_2 \end{bmatrix} = \begin{bmatrix} b_{feas} \\ b_{temp} \end{bmatrix}$$
$$[\Phi, \xi_1 \xi_2] \in -P$$

Note that the hypotheses of Theorem 4.2.3 are satisfied. The operator representing the equality constraints has closed range, since $\mathcal{A}_{feas}$ does. It is easy to verify that the positive cone P has a nonempty interior and the regularity condition is satisfied by assumption.

We can readily write the dual of the above problem according with Theorem 4.2.3, where

$$\lambda \in P_1^{\oplus} + P_2^{\oplus} = \{\Phi^*, \xi_1^*, \xi_2^* \mid \|\Phi^*\|_\infty + \xi_1^* \leq 0, \; \xi_1^* \leq 0, \; \xi_2^* \leq 0\}.$$

It is left to the reader to verify that, after some rearrangements, the resulting dual problem is given by:

$$\max \quad \langle b_{feas}, z_1^* \rangle + \langle b_{temp}, z_2^* \rangle$$
$$\text{subject to:}$$
$$\|\mathcal{A}_{feas}^* z_1^* + \mathcal{A}_{temp}^* z_2^* \|_\infty \leq 1 \tag{5.2}$$
$$z_2^* \leq 0$$
$$z_1^* \in \ell_\infty, \; z_2^* \in \mathbb{R}^n,$$

which is the Lagrangian dual, with no duality gap, of Problem (5.1). ■

Remark 5.2.1. The assumption on the regularity condition does not pose any problem in practice. If, for a given b_{temp}, Problem (5.1) is feasible, then, it is still feasible when b_{temp} is replaced by $b_{temp} + 1\delta$, $\delta > 0$, and the problem satisfies the regularity condition with $\epsilon < \delta$.

The decay of the rows of $A_{temp} = \mathcal{A}_{temp}\mathcal{O}^*$, the matrix representation of $\mathcal{A}_{temp}$, makes this problem very similar to the standard ℓ_1 problem.

Definition 5.2.1. *The vector space of finite support sequences of matrices, denoted by $FS^{m \times n}$, is the set of all the sequences of real $m \times n$ matrices with only a finite number of nonzero matrices.*

Notice that a feasible closed loop map, Φ, has a Finite Impulse Response if and only if $\Phi \in FS^{n_z \times n_w}$.

Corollary 5.2.1. *In the one-block case, Problem (5.1) is equivalent to a finite dimensional linear programming problem, and the optimal solution has Finite Impulse Response.*

Proof. The argument is similar to the one used in the standard ℓ_1 problem and is possible to extend because of the duality representation in (5.2). Here we just sketch it. One-block problems have only zero interpolation conditions and no rank interpolation conditions, i.e., $\mathcal{A}_{feas} = \mathcal{A}_Z$ has range in $\mathbb{R}^{c_z}$ and the rows of the matrix representation of the operators $\mathcal{A}_{feas}$ and $\mathcal{A}_{temp}$ are elements of c_0.

From these properties it follows that the dual problem has only a finite number $(c_z + T)$ of variables and, although there is an infinite number of constraints, the number of dual constraints that are active is limited, since the columns of A^T_{feas} and A^T_{temp} decay. From the alignment conditions, this implies that the optimal solution for Problem (5.1) is a finite support sequence and it can be computed exactly by solving a finite dimensional linear programming problem. ■

In the multi-block case, the dual problem lies in $\ell_\infty \times \mathbb{R}^n$ and in general it is not equivalent to a finite dimensional linear programming problem. However, the problem at hand is also the dual, with no duality gap, of an optimization problem, the (pre-)dual, posed in $c_0 \times \mathbb{R}^n$. Corollary 4.2.1 is the natural candidate to prove the above statement. To apply it, however, we need $\mathcal{A}_{feas}$ and $\mathcal{A}_{temp}$ to have a pre-adjoint (Def. 2.4.3).

Since $\mathcal{A}_{temp}$ has range in $\mathbb{R}^n$, we have that ${}^*\mathcal{A}_{temp} = \mathcal{A}^*_{temp}$. From [15] it also follows that ${}^*\mathcal{A}_{feas} : c_0 \to c_0^{n_z \times n_w}$ exists and is equal $\mathcal{A}^*_{feas}|_{c_0}$, i.e., it is given by the adjoint operator with domain restricted to c_0.

Theorem 5.2.2. *Assume that Problem (5.1) is feasible, then, under the current assumptions, the optimal minimizing solution for (5.1) exists and the optimal cost μ^o is given by:*

$$\begin{aligned} \mu^o = \quad & \sup \quad \langle b_{feas}, z_1\rangle + \langle b_{temp}, z_2\rangle \\ \textit{subject to:} \quad & \\ & \|{}^*\mathcal{A}_{feas} z_1 + {}^*\mathcal{A}_{temp} z_2\|_\infty \le 1 \\ & z_2 \le 0 \\ & z_1 \in c_0,\ z_2 \in \mathbb{R}^n \end{aligned} \tag{5.3}$$

Proof. The optimal value of (5.3) is finite, since it is smaller than the optimal value of (5.2) and the latter is always smaller than or equal to μ^o. We can rewrite Problem (5.3) as follows:

$$\begin{aligned} \nu^o = \quad & \sup \quad \langle b_{feas}, z_1\rangle + \langle b_{temp}, z_2\rangle \\ \text{subject to:} \quad & \\ & {}^*\bar{A} z + c \ge 0 \end{aligned}$$

where ${}^*\bar{A} = \begin{bmatrix} {}^*\mathcal{A}_{feas} & {}^*\mathcal{A}_{temp} \\ 0 & 0 \\ 0 & I \end{bmatrix}$ $z = \begin{bmatrix} z_1 \\ z_2 \end{bmatrix}$, $c = \begin{bmatrix} 0 \\ 1 \\ 0 \end{bmatrix}$, and the inequality is defined with respect to the following cone in $c_0^{n_z \times n_w} \times \mathbb{R} \times \mathbb{R}^n$:

$$P = \{w, \gamma, \xi \mid \|w\|_\infty - \gamma \le 0, \gamma \ge 0, \xi \le 0\}.$$

Notice that P has a nonempty interior. Moreover, a feasible solution in the interior of P can be found as follows: fix $z_1 = 0$, since $\mathcal{A}_{temp}$ is a bounded linear operator $\mathcal{A}^*_{temp}$ is bounded and linear too. Thus there is a number denoted as $\|\mathcal{A}^*_{temp}\|$ such that $\|\mathcal{A}^*_{temp} z_2^*\|_\infty \leq \|\mathcal{A}^*_{temp}\| \|z_2^*\|_\infty$. Then, pick any $z_2 < 0$ such that $\|z_2\|_\infty \leq \epsilon$ with $0 < \epsilon < \frac{1}{\|\mathcal{A}^*_{temp}\|}$. Hence, Corollary 4.2.1 applies and the result follows from simple manipulations. ■

Remark 5.2.2. This theorem has important implications. First of all, a minimizing optimal solution to Problem (5.1) always exists. Second, since the set of dual feasible solutions, $z = [z_1, z_2]$, with finite support is dense in the set of dual feasible solutions in $c_0 \times \mathbb{R}^n$, the optimal cost of the (pre-)dual problem can be approximated arbitrarily well by truncating the dual variables. This provides a computable lower bound for the optimal cost, since the resulting truncated problems are finite dimensional linear programs.

Notice that, since there is no duality gap between the (pre-)dual problem and the primal, then, there is no gap between the primal and the dual. This follows from the fact that the cost of the dual problem is always greater than the cost of the (pre-)dual problem and smaller than the cost of the primal problem. Based on the above property, it is possible to show that the sequence of finite support solutions converges *weak** to the optimal dual solution. We will not present this result.

5.3 ℓ_1 Control with Infinite Horizon Time-domain Constraints

Here we study the optimal ℓ_1 problem with infinite horizon time domain constraints on the response to a fixed input $w \in \ell_\infty^{n_w}$. There are several issues that make the study and the solution to this problem more complicated than in the finite horizon case. For the sake of simplicity, we assume that only a single output, say z_1, has to be constrained. The generalization to the multi-output case is straightforward but tedious and therefore is omitted.

For a fixed input $w \in \ell_\infty^{n_w}$, generally, $\mathcal{A}_{temp}$ maps $\ell_1^{n_z \times n_w}$ to ℓ_∞. The natural positive cone in ℓ_∞ does have a nonempty interior. Assuming that the regularity condition is satisfied, then Theorem 4.2.2 applies and establishes the lack of duality gap between the primal and the dual problem. In this case, the variables of the dual problem lie in $\ell_\infty \times {\ell_\infty}^*$. Unfortunately, the action of a linear functional z^* in ${\ell_\infty}^*$ on elements z in ℓ_∞ cannot be solely represented in terms of scalar product between z^* and z, i.e., not all linear functionals on ℓ_∞ act as $\langle z, z^* \rangle = \sum_{i=1}^{\infty} z^*(i) z(i)$. Thus, although it is possible to establish no duality gap between the primal problem:

$$\mu^o = \inf_{\substack{\Phi \in \ell_1^{n_z \times n_w} \\ \mathcal{A}_{feas}\Phi = b_{feas} \\ \mathcal{A}_{temp}\Phi \leq b_{temp}}} \|\Phi\|_1, \tag{5.4}$$

with range of $\mathcal{A}_{temp}$ in ℓ_∞, and its dual:

$$\max_{\substack{\|\mathcal{A}_{feas}^* z_1^* + \mathcal{A}_{temp}^* z_2^*\|_\infty \leq 1 \\ z_1^* \in \ell_\infty,\ z_2^* \leq 0,\ z_2^* \in \ell_\infty{}^*}} \langle b_{feas}, z_1^* \rangle + \langle b_{temp}, z_2^* \rangle, \tag{5.5}$$

the dual problem is defined only abstractly, since we don't have a concrete representation for all the elements in $\ell_\infty{}^*$ and hence for $\mathcal{A}_{temp}^*$.

On the other hand, if we restrict our attention to input signals that give rise to responses in c_0, then, although we now have a representation for the dual, the natural positive cone in c_0 has empty interior. In this situation, the possible lack of duality gap cannot be established through Theorem 4.2.2, since one of the hypotheses is not satisfied.

Thus, c_0 is too small and ℓ_∞ is too large for our purposes. Consider the vector space c defined as follows:

Definition 5.3.1. *The space c is given by the set of all sequences in ℓ_∞ whose limit exists and is finite.*

Notice that c_0 is a closed subspace of c, and c is a closed subspace of ℓ_∞ and therefore it is a Banach space. Any element x in c is uniquely given by the sum of an element in c_0 and a step sequence, i.e., $c = c_0 \oplus s$, where $s = \{w \in \ell_\infty \,|w = \alpha w_s,\ \alpha \in \mathbb{R},\ w_s = \text{unit step sequence}\}$.

The choice of c as underlying space is more natural, since the step response of a stable system is a sequence in c. Moreover, the positive cone in c has a non empty interior, and the operator $\mathcal{A}_{temp} : \ell_1^{n_z \times n_w} \to c$ has closed range. Thus, it is possible to achieve a representation of the dual problem with no duality gap in $\ell_\infty \times \ell_1$ since the dual of c is isometrically isomorphic to ℓ_1. Without going into details, this is saying that, with a slight abuse of notation, we can replace $z_2 \in \ell_\infty{}^*$ in Problem 5.5 with $z_2 \in \ell_1$.

The above discussion highlights that a careful selection of the space where the problem is posed is important in obtaining useful results from the theory presented in this monograph.

Next Theorem studies the possibility of approximating the optimal cost of the dual problem by solving finite dimensional optimizations. It states that this approximation is possible in the space of finite support sequences. Define the following problem:

$$\nu^o = \sup_{\substack{\|\mathcal{A}_{feas}^* z_1^* + \mathcal{A}_{temp}^* z_2^*\|_\infty \leq 1 \\ z_1^* \in FS,\ z_2^* \leq 0 \in FS}} \langle b_{feas}, z_1^* \rangle + \langle b_{temp}, z_2^* \rangle, \tag{5.6}$$

Problem (5.6) is nothing but Problem (5.5), dual of Problem (5.4), with the variables restricted to be elements of $FS \times FS$. We are going to show that,

Problem (5.6) and Problem (5.5) have the same optimal cost. Thus Problem (5.6) provides a way to compute lower bounds of μ^o within any accuracy.

Theorem 5.3.1. *Assume that Problem (5.4) is feasible, and that b_{temp} belongs to the space c. Then an optimal solution Φ^o exists in the primal space, and the optimal cost, μ^o, is equal to the optimal cost, ν^o, of Problem (5.6).*

Proof. The detailed proof is in the Appendix. Here we outline the main steps. We consider the following sequence problems with truncated horizon:

$$\mu_N^o = \inf_{\Phi \in \ell_1^{n_z \times n_w}} \|\Phi\|_1 \tag{5.7}$$

subject to:

$$\mathcal{A}_{feas}\Phi = b_{feas}$$
$$P_N(\mathcal{A}_{temp})\Phi \leq P_N(b_{temp})$$

where $P_N : \ell_\infty \to \mathbb{R}^N$ is a truncation operator. From Theorem 5.2.2 and Remark 5.2.2, we know that an optimal primal solution exists and μ_N is also given by the pre-dual problem with $z_1^* \in FS$ and $z_2^* \in \mathbb{R}^N$. Then, we show that μ_N converges to μ^o by showing that the sequence of primal solutions Φ_N^o converges *weak** an optimal solution to the infinite horizon problem. ■

5.3.1 Approximation of the Primal

Until now, we have considered the approximation of the dual problem. We have shown that a lower bound for the optimal cost can be computed by forcing the dual variables to have finite support, and that the lower bound converges to the optimal cost as the support of the dual solution increases. This computational method is known as FME in the ℓ_1 literature [16, 15, 1], and we have shown that it is directly applicable to the ℓ_1 control problem with time-domain constraints.

We can think of applying the same idea to the primal problem, forcing the primal variables to have finite support. This method is known as FMV in the ℓ_1 literature [16, 15, 1]. To prove that it is possible to approximate the primal problem, one has to show the denseness of the finite support feasible solutions in the set of all feasible solutions. Not all problems satisfy this condition. For example, even in the standard ℓ_1 problem ([16]), there is a class of problems with no FIR feasible solutions. However, from the denseness of finite support sequences in ℓ_1 and the Youla parametrization, it follows that if there is an achievable $\Phi = H - UQV$ with finite support for some $Q \in \ell_1^{n_z \times n_w}$, then the achievable FIR closed loop systems are dense in the set of all achievable closed loop systems ([14]).

Next Theorem generalizes the result in [14] and shows that if there is a FIR feasible solution, Φ, to Problem (5.4) in the interior of the cone that defines the inequality constraints, i.e., $\mathcal{A}_{temp}\Phi < b_{temp}$, then, the FIR feasible solutions are dense in the feasible set. For the sake of simplicity we prove the theorem in the special case where the set of all achievable closed loop maps is given by $\Phi = H - UQ$, $Q \in \ell_1^{n_u \times n_w}$. The proof in the general case follows the same argument but a more complicated notation is needed. We would like to point out that the argument in the proof relies on duality theory.

Theorem 5.3.2. *Consider Problem (5.4). Assume that* $b_{temp} \in c$, *and that there exists a feasible solution* Φ_p *with finite support such that* $\mathcal{A}_{feas}\Phi_p = b_{feas}$, *and* $\mathcal{A}_{temp}\Phi_p < b_{temp}$. *Then*

$$\mu^o = \inf_{\Phi \in FS^{n_z \times n_w}} \|\Phi\|_1 \tag{5.8}$$

subject to:

$$\mathcal{A}_{feas}\Phi = b_{feas}$$
$$\mathcal{A}_{temp}\Phi \leq b_{temp}$$

Proof. See Appendix. ∎

Thus, under the hypothesis of the above theorem, Problem (5.4) can be posed as a minimization in the space of finite support sequences. The FMV method can be applied to obtain suboptimal feasible FIR solutions, whose norm converges to the optimal cost as the length of the support increases.

5.4 Example

In this section, we compute the solution to an optimal ℓ_1 control problem with time domain constraints on the response to a step input. We use the FME and FMV methods previously described. These methods are based on computing, respectively, FIR sub-optimal and FIR super-optimal solutions to the given problem. This is achieved by forcing respectively the support of the primal variables to be finite and the support of the dual variables to be finite. From the FMV method we obtain a sub-optimal FIR feasible solution to the problem at hand, while from the FME method we obtain a lower bound on the optimal cost useful to compute the accuracy of the FIR solution.

The problem here considered satisfies the hypotheses of Theorem 5.3.2, thus, an approximation of the primal problem with a FIR feasible solution is possible in this case.

By solving the problem for different values of a performance parameter, we obtain a quantitative description of the tradeoffs and of the fundamental limitations of the closed loop map.

Consider the system described in Figure 5.1, where the discrete-time plant P has the following $\lambda - transform$:

$$P(\lambda) = \frac{\lambda^2}{1 - .5\lambda + 2.5\lambda^2}$$

and the weighting function W is given by:

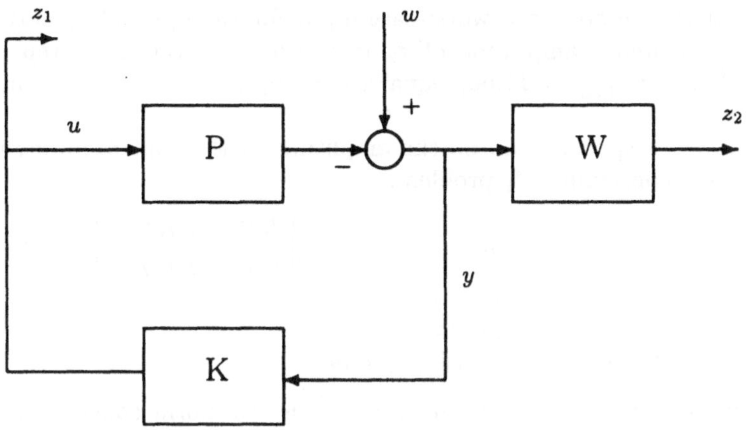

Fig. 5.1. System Configuration

$$W(\lambda) = \frac{1.5}{1 - .7\lambda}$$

Note that the plant has two unstable poles at $0.25 \pm j1.5612$ and two zeros at the origin. It is well known that such plants are particularly difficult to control.

Besides stability, the general objective is to reject unknown but bounded amplitude disturbance, while limiting the worst-case control input amplitude. Moreover, it is desired that the amplitude of the control input does not exceed the normalized bound of 4 units when the disturbance is a unit step. The weight W is used to emphasize the disturbance rejection at low frequencies.

Given the nature of the disturbance, our first attempt is to design an ℓ_1 optimal controller. Note that the problem is multi-block. In particular, it is a two block problem in which the ℓ_1 norm from w to z_1 and z_2 is minimized. The computation of $\mathcal{A}_{feas}$ is cumbersome to be carried out by hand. The feasibility constraints are automatically computed by a computer program following the characterization given in [16].

The resulting optimal ℓ_1 norm of the closed loop system is $\mu^{\circ}_{\ell_1} = 11.66$. The optimal closed loop transfer function matrix is FIR and is given by:

$$\hat{\Phi}_1(\lambda) = 1.8012 + 0.3000\lambda + 2.9174\lambda^2 + 2.8096\lambda^3 - 2.1207\lambda^4 - 1.7113\lambda^5$$
$$\hat{\Phi}_2(\lambda) = 1.5 + 1.05\lambda + 3.4367\lambda^2 + 4.2066\lambda^3 + 1.4669\lambda^4$$

The optimal controller is of order 3 and is given by:

$$\hat{K}(\lambda) = \frac{1.8012 + 1.2006\lambda - 0.9852\lambda^2 - 0.6845\lambda^3}{1 + 0.5\lambda - 0.4488\lambda^2 - 0.2738\lambda^3}$$

Using the optimal controller we can verify that the maximum control input amplitude in response to a unit step disturbance, w_s, is about 7.9 (see Figure 5.3), and violates the constraint of being less than 4.

Note that, the step is a worst-case input for the optimal ℓ_1 closed loop, since the maximum amplitude of z_2 in response to the unit step input is $\|W(1-PK)^{-1}w_s\|_\infty = 11.66$, equal to the optimal ℓ_1 norm of the closed loop system.

To meet the specification on the amplitude of the control input, we pose the following constrained ℓ_1 problem:

$$\inf \left\| \begin{matrix} K(1-PK)^{-1} \\ W(1-PK)^{-1} \end{matrix} \right\|_1$$

$$\begin{gathered} \text{subject to:} \\ K - \textit{stabilizing} \\ \|K(1-KP)^{-1}w_s\|_\infty \leq umax \end{gathered}$$

where $umax = 4$ in this case. In practice, the ℓ_∞ norm constraints are imposed only on the first 30 samples of the control input response. The extra constraints are easily added to the linear program derived for the ℓ_1 problem. By simulation, it is possible to verify that the transient dies out well before the 30th sample, making the ℓ_∞ constraints on the samples after the 30th inactive in the optimization. This also implies that the amplitude constraint is satisfied for the infinite-horizon problem.

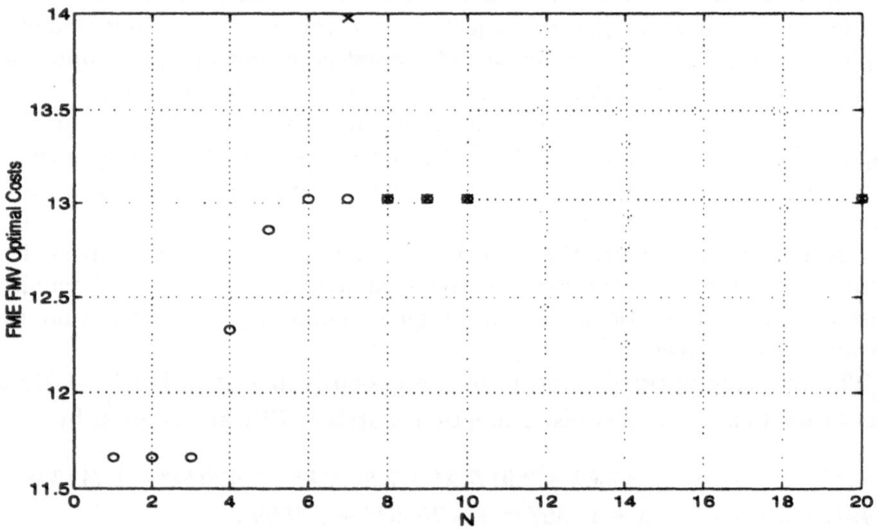

Fig. 5.2. Convergence of the FMV costs (x) and of the FME costs (o) to the optimal ℓ_1 cost for umax=4.

Figure 5.2 shows the convergence of the upper and lower bounds given by the solution to FMV and FME problems for various values of N (the order of approximation). N is the number of variables allowed to be different from zero in the FMV problem and is the number of dual variables allowed to be different from zero in the FME problem. Note that FMV has no feasible solution for $N < 7$. For $N = 8$ the FMV solution has converged to the cost $\mu^o = 13.02$ which is the optimal cost of the problem since also the FME costs converge to it. The convergence of the FME costs is more gradual. The lower bound converges to μ^o for $N = 6$. The upper and lower bounds are the same for $N \geq 8$, which is a clear indication that the optimal ℓ_1 solution is an FIR of order 7 and is captured by both methods.

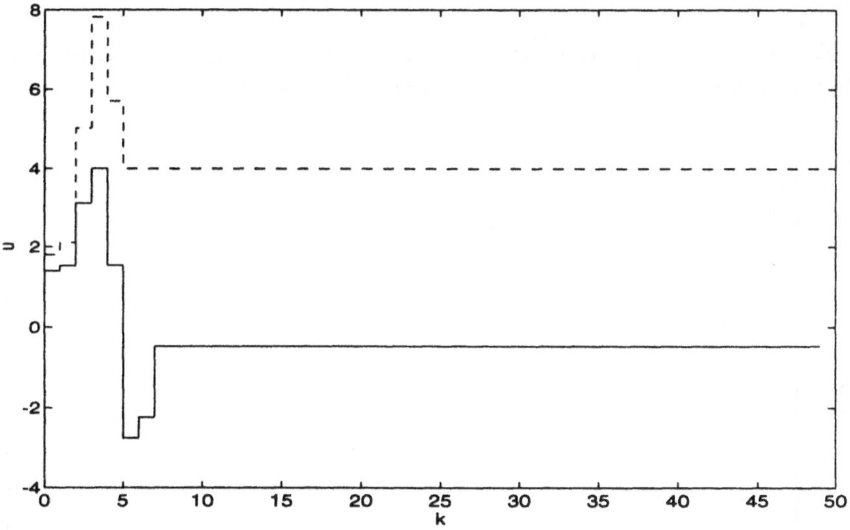

Fig. 5.3. Closed Loop Control Input Response to a Unit Step Input, constrained with umax=4 (solid-line), unconstrained (dashed-line)

Figure 5.3 shows the closed loop responses of $u = z_1$, due to a unit step input in the constrained and unconstrained case. The optimal Φ in the constrained case in given by:

$$\begin{aligned}\hat{\Phi}_1(\lambda) &= 1.4006 + 0.1274\lambda + 1.5818\lambda^2 + 0.8901\lambda^3 - 2.4542\lambda^4 - 4.2968\lambda^5 \\ &\quad +0.5079\lambda^6 + 1.7644\lambda^7 \\ \hat{\Phi}_2(\lambda) &= 1.5 + 1.05\lambda + 2.8359\lambda^2 + 3.2268\lambda^3 - 2.8982\lambda^5 - 1.5123\lambda^6\end{aligned}$$

The resulting optimal controller is of order 5 and given by:

$$\hat{K}(\lambda) = \frac{1.4006 + 0.8277\lambda - 1.5058\lambda^2 - 1.9322\lambda^3 + 0.3443\lambda^4 + 0.7057\lambda^5}{1 + 0.5\lambda - 0.8494\lambda^2 - 0.8469\lambda^3 + 0.1942\lambda^4 + 0.2823\lambda^5}$$

5.4.1 Fundamental Closed Loop Limitations and Performance Tradeoffs

We can use the technique proposed in this monograph to find out limitations of the feedback system due to the plant characteristics. Some general results on the frequency domain intrinsic limitations of the closed loop system are given in [33]. The results for time domain performance are mostly qualitative. Using the technique discussed in the monograph, we are able to quantitatively characterize the fundamental limitations of the closed loop map.

In the previous part, we were able to find a stabilizing controller that minimizes the ℓ_1 norm of the closed loop system and uses a control input of maximum peak amplitude less than 4 units, when the input signal is a unit step. Now, we would like to know how small the maximum amplitude of the control input can be to still ensure the stability of the nominal closed loop system.

To answer this question, we solve the constrained ℓ_1 problem for various values of $umax$. The results are reported in Figure 5.4.

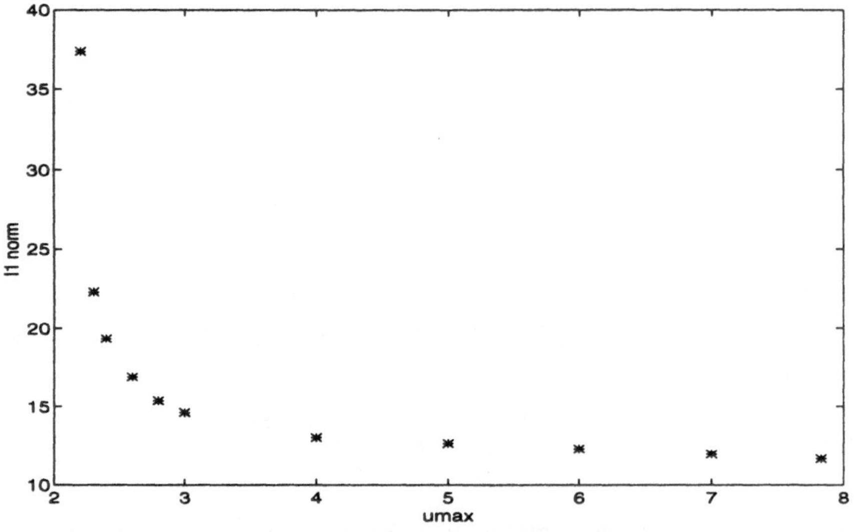

Fig. 5.4. Performance Tradeoffs

The graph of the ℓ_1 optimal norm as function of $umax$ has a vertical asymptote for $umax \approx 2.19$. The result states that it is impossible to stabilize P with a linear controller that uses a control input of amplitude less the 2.19 units for a unit step disturbance input. Note that the curve is pretty flat for $umax \geq 4$. This indicates that the maximum amplitude of the control input can be considerably reduced with a minor increment in the worst-case performance. It is possible to reduce $umax$ of about 50% with a 12% increment of the ℓ_1 norm.

Other useful information can be obtained by comparing the maximum peak output due to a step input both for the unconstrained case and for the case $umax = 4$. As already mentioned, in the unconstrained case the step input is a worst-case input for the optimal closed loop system. For $umax = 4$, the step is no longer the worst-case input, since $\|W(1-PK)^{-1}w_s\|_\infty = 8.61$ and $\|K(1-PK)^{-1}w_s\|_\infty = 4$, while $\|\Phi\|_1 = 13.02$. Moreover, the overall maximum peak output due to a step input is improved compared to the unconstrained case, since $\|\Phi w_s\|_\infty = 8.61$, while, in the unconstrained case, $\|\Phi w_s\|_\infty = 11.66$. Also, in this case, the tradeoffs between worst-case performance and special-case performance is readily stated in quantitative terms.

Summary and Comments. In this chapter we have generalized most of the results developed for the standard ℓ_1 problem to the ℓ_1 control problem with time-domain constraints. The study has been almost exhaustive. We have derived the dual problem, established the lack of duality gap, determined the existence of a minimizing solution, proved that one-block problems with finite-horizon constraints have FIR optimal solutions, and shown that the computational methods, based on finite support approximation developed for the ℓ_1 problem, retain their properties when applied to this multi-objective problem.

6. ℓ_1 -Minimization with Magnitude Constraints in the Frequency Domain

In this chapter, we consider the ℓ_1 optimal control problem with constraints on the magnitude of the closed loop frequency response. In particular, we consider the case where the constraints are imposed at discrete points of the frequency response, since a finite number of them may be sufficient to shape the closed loop frequency response.

This problem often arises in multi-objective control system design, where the controller has to be designed so that the closed loop system satisfies several performance specifications both in the time and in the frequency domain. In many cases, only small adjustments in limited regions of the response are needed to meet the specifications. When an automatic design methodology based on optimal control is used to carry out the design, these adjustments translate into fewer constraints to be imposed in the optimization and therefore it yields to a reduction in the computational complexity of the problem to be solved. For example, a finite number of magnitude constraints at fixed frequency points could be used to approximate an $\mathcal{H}_\infty$ constraint on the closed loop map. The approach we take for this problem is compatible with the methods of Chapter 5. Thus, we can add both time-domain constraints and frequency constraints to the standard ℓ_1 problem, and obtain computable upper and lower bounds to the optimal cost.

The main point that we make in this chapter is that the lack of duality gap in a infinite dimensional problem does not implies that we can compute converging primal and dual finite dimensional approximations. Here, we show that, for multi-block problems, finite support feasible dual solutions may not approximate arbitrarily well the optimal cost.

We formalize the problem by representing the frequency point magnitude constraints as an uncountable set of linear constraints, and derive the resulting primal and dual formulations in terms of infinite dimensional optimization problems. We also derive the equivalence of primal and dual problem with LMI problems.

Through the analysis of the structure of the dual problem, we study the approximation of the convex constraints representing the frequency domain magnitude constraints, with a finite number of linear constraints. We show that the LMI problem can be approximated arbitrarily well by a linear programming problem.

Whether in terms of LMI or in terms of approximate LP, the resulting problem may still be infinite dimensional. In these cases, we need to study finite dimensional problems that provide computable approximations of the optimal solution to the infinite dimensional problem at hand. In particular, we study the applicability of the approximation methods introduced in Chapter 5, based on approximating the primal and the dual space with the space of finite support sequences.

While the convergence of the primal approximation method to the optimal cost is guaranteed, the situation for the dual approximation method is not as nice. We show, using duality theory, that this method is not guaranteed to converge to the optimal cost. In particular, we present an example in which there is a gap between the optimal cost of the problem at hand and the cost of the best dual approximation of finite support. Finding converging lower bounds is an open area of research. A possible approximation method is suggested at the end of the chapter.

Other approaches to the mixed $\ell_1/\mathcal{H}_\infty$ problem have been considered in [12, 34].

6.1 Problem Statement

Suppose that in a MIMO problem we want to impose some frequency domain constraints on some of the single SISO transfer functions of the closed loop map. For the sake of simplicity, we will consider the case where only one frequency point constraint is imposed on one of the SISO transfer functions of the MIMO closed loop map Φ. The case of finite number of frequency point constraints imposed on different SISO transfer functions of Φ is a straightforward extension.

Without loss of generality we assume that the frequency point magnitude constraint has to be imposed on Φ_{11}.
The problem we want to solve can be expressed as follows:

$$\mu^o = \inf_{\Phi \in \ell_1^{n_z \times n_w}} \|\Phi\|_1 \tag{6.1}$$

subject to:

$$\mathcal{A}_{feas}\Phi = b_{feas}$$
$$|\hat{\Phi}_{11}(e^{j\omega})| \leq \gamma$$

There are several ways to obtain a dual for this problem. One is to represent the magnitude constraint by the LMI in Section 3.3.4 Equations (3.7), (3.8) and (3.9). The theory of Chapter 4 can be applied using the representation of Section 4.1. The resulting dual problem can be derived from the the general case presented in Section 4.2.1, and it can also be found in [21]. Incidentally, the way the dual in [21] was derived is not the one just described but by extending approach we follow here based on representing the LMI constraints as uncountably infinite set of linear constraints.

Another possibility is to represent the magnitude constraint in terms of the real and imaginary part of the discrete Fourier transform, i.e.,

$$|\hat{\Phi}_{11}(e^{j\omega})| \leq \gamma \Longleftrightarrow \sqrt{\Phi_{11,R}(\omega)^2 + \Phi_{11,I}(\omega)^2} \leq \gamma$$

where

$$\Phi_{11,R}(\omega) = Re\hat{\Phi}_{11}(e^{i\omega}) = \sum_{n=0}^{\infty} \Phi_{11}(n)\cos(n\omega)$$

$$\Phi_{11,I}(\omega) = Im\hat{\Phi}_{11}(e^{i\omega}) = \sum_{n=0}^{\infty} \Phi_{11}(n)\sin(n\omega).$$

If we denote by $F : \ell_1 \to \mathbb{R}^2$ the following linear operator:

$$F = \begin{bmatrix} 1 & \cos(\omega) & \cdots & \cos(n\omega) & \cdots \\ 0 & \sin(\omega) & \cdots & \sin(n\omega) & \cdots \end{bmatrix}$$

then the magnitude constraint can be rewritten as follows

$$\begin{bmatrix} I & -\mathcal{F} & 0 \\ 0 & 0 & 1 \end{bmatrix} \begin{bmatrix} y \\ \Phi \\ \xi \end{bmatrix} = \begin{bmatrix} 0 \\ \gamma \end{bmatrix}$$

where $\mathcal{F}\Phi = F\Phi_{11}$, and y, Φ, ξ belong to the cone

$$P = \{(y, \Phi, \xi) \in \mathbb{R}^2 \times \ell_1 \times \mathbb{R} \mid \|y\|_2 \leq \xi,\ \|\Phi\|_1 \leq \mu,\ \xi \geq 0\}$$

With this definition, we could directly proceed and apply Theorem 4.2.3 to obtain the following dual problem with no duality gap.

$$\max_{z_1^*, C, \eta} \quad \langle b_{feas}, z_1^* \rangle - \gamma\eta$$

subject to:

$$\begin{aligned} &\|C\|_2 \leq \eta \\ &\|\mathcal{A}_{feas}^* z_1^* + \mathcal{F}^* \begin{bmatrix} C_1 \\ C_2 \end{bmatrix}\|_{\infty} \leq 1 \\ &z_1^* \in \ell_\infty,\ C \in \mathbb{R}^2,\ \eta \geq 0 \end{aligned} \tag{6.2}$$

where

$$[\mathcal{A}^* C]_{ij} = \begin{cases} F^T C, & i = j = 1, \\ 0, & i = 2, \ldots, n_z\ j = 2, \ldots, n_w, \end{cases}$$

In this case, Theorem 4.2.3 will guarantee the lack of a duality gap between two convex optimization problems. Since, one of the topic of this chapter is to study the approximation of these convex constraints by a finite set of linear constraints, a more convenient representation of the magnitude constraint is in terms of an infinite number of linear constraints.

A magnitude constraint on $\hat{\Phi}_{11}(\omega)$, the discrete Fourier transform of $\Phi_{11} \in \ell_1$ at some frequency ω, is equivalent to an infinite number of linear constraints on Φ_{11} as follows:

$$|\hat{\Phi}_{11}(\omega)| \leq \gamma \Longleftrightarrow \Phi_{11,R}(\omega)\cos\theta + \Phi_{11,I}(\omega)\sin\theta \leq \gamma, \quad \forall\theta \in [0, 2\pi). \quad (6.3)$$

Notice that these are indeed an uncountable number of constraints.

Using the definition of discrete Fourier transform, the above constraint can be represented as $\mathcal{A}_{\mathcal{H}_\infty}\Phi \leq \gamma$, with $\mathcal{A}_{\mathcal{H}_\infty} : \ell_1^{n_z \times n_w} \to C_{[0,2\pi)}$ such that

$$\mathcal{A}_{\mathcal{H}_\infty}\Phi = \sum_{n=0}^{\infty} \Phi_{11}(n)(\cos(\theta)\cos(n\omega) - \sin(\theta)\sin(n\omega)) \qquad \theta \in [0, 2\pi). \quad (6.4)$$

We now provide a characterization of $\mathcal{A}^*_{\mathcal{H}_\infty}$.

The dual space of $C_{[0,2\pi)}$ is the space BV of functions of bounded variations on $[0, 2\pi)$ [22]. Thus, the adjoint of $\mathcal{A}_{\mathcal{H}_\infty}$ maps BV into $\ell_\infty^{n_z \times n_w}$.
Consider the adjoint of the operator F. It follows that $F^* : \mathbb{R}^2 \to \ell_\infty$ is just the transpose of F.

$$F^* = \begin{bmatrix} 1 & 0 \\ \cos(\omega) & \sin(\omega) \\ \vdots & \vdots \\ \cos(n\omega) & \sin(n\omega) \\ \vdots & \vdots \end{bmatrix}$$

If we let:

$$C_1(\rho) = \int_0^{2\pi} \cos(\beta)d\rho(\beta), \quad C_2(\rho) = \int_0^{2\pi} \sin(\beta)d\rho(\beta) \quad (6.5)$$

then, $\mathcal{A}^*_{\mathcal{H}_\infty}$ is characterized as follows:
for any $\rho(\theta) \in BV$ the matrix of sequences $\mathcal{A}^*_{\mathcal{H}_\infty}\rho \in \ell_\infty^{n_z \times n_w}$ is given by:

$$\left[\mathcal{A}^*_{\mathcal{H}_\infty}\rho\right]_{ij} = \begin{cases} F^* \begin{bmatrix} C_1(\rho) \\ C_2(\rho) \end{bmatrix} & i = j = 1 \\ 0 & , \ i = 2, \ldots, n_z \ j = 2, \ldots, n_w \end{cases} \quad (6.6)$$

where 0 denotes the zero sequence.

6.2 Primal-Dual Formulation

In this section, we derive the primal-dual formulation, with no duality gap, in terms of the infinite linear constraints representation of the magnitude constraint. The problem we want to study can be formulated as follows:

$$\mu^o = \inf_{\Phi \in \ell_1^{n_z \times n_w}} \|\Phi\|_1 \quad (6.7)$$

subject to:

$$\mathcal{A}_{feas}\Phi = b_{feas}$$
$$\mathcal{A}_{\mathcal{H}_\infty}\Phi \leq \gamma$$

where $\mathcal{A}_{\mathcal{H}_\infty}$ is given in Equation (6.4). Note that for each $\theta \in [0, 2\pi)$, there is a linear inequality constraint on Φ, thus, the above is a linear programming problem. Next theorem gives a characterization of the dual problem with no duality gap.

Theorem 6.2.1. *Under the current assumptions, suppose that problem (6.7) is feasible and there exists an $\epsilon > 0$ such that, if γ is replaced by $\gamma - \epsilon$, problem (6.7) stays feasible. Then, the dual of Problem (6.7) is given by:*

$$\max_{z_1^*, C} \quad \langle b_{feas}, z_1^* \rangle - \gamma\alpha$$

subject to:

$$\left\| \mathcal{A}_{feas}^* z_1^* + \bar{\mathcal{A}}_{\mathcal{H}_\infty}^* \alpha \begin{bmatrix} \cos(\theta) \\ \sin(\theta) \end{bmatrix} \right\|_\infty \leq 1, \tag{6.8}$$

$$z_1^* \in \ell_\infty, \ \alpha \geq 0, 0 \leq \theta \leq 2\pi$$

where

$$\left[\bar{\mathcal{A}}_{\mathcal{H}_\infty}^* \alpha \begin{bmatrix} \cos(\theta) \\ \sin(\theta) \end{bmatrix} \right]_{ij} = \begin{cases} F^* \alpha \begin{bmatrix} \cos(\theta) \\ \sin(\theta) \end{bmatrix}, & i = j = 1, \\ 0, & i = 2, \ldots, n_z \ j = 2, \ldots, n_w, \end{cases}$$

and there is no duality gap.

Proof. See Appendix. ∎

Notice that this is nothing but Problem 6.2 in terms of an infinite set of linear constraints. To see the equivalence let $C = \alpha \begin{bmatrix} \cos(\theta) \\ \sin(\theta) \end{bmatrix}$, and note that $\bar{\mathcal{A}}_{\mathcal{H}_\infty} = \mathcal{F}$.

Remark 6.2.1. We must remark that, for the problem under study, the regularity condition holds for all $\gamma > 0$. The magnitude constraint is equivalent to the following: $\hat{\Phi}(e^{j\omega})\hat{\Phi}(e^{-j\omega}) \leq \gamma^2$. Recall that any Φ such that $\mathcal{A}_{feas}\Phi = b_{feas}$ is given by $\Phi = H - UQV$. Since we are imposing a frequency constraint on Φ_{11}, denote by $\hat{U}_1^T$ and $\hat{V}_1$ the first row of $\hat{U}$ and the first column of $\hat{V}$. Note that $\hat{\Phi}_{11} = \hat{H}_{11} - \hat{U}_1^T \hat{Q} \hat{V}_1$. Under the current assumptions, it follows that the $\hat{U}_1^T$ and $\hat{V}_1$ have respectively full row rank and full column rank on the unit circle. Thus we can always impose that $\hat{\Phi}_{11}(e^{j\omega}) = \xi < \gamma$, and hence $\hat{\Phi}(e^{j\omega})\hat{\Phi}(e^{-j\omega}) < \gamma^2$, by selecting any Q such that

$$\hat{U}_1^T(e^{j\omega})\hat{Q}(e^{j\omega})\hat{V}_1(e^{j\omega}) = \hat{H}(e^{j\omega}) - \xi.$$

Comment: We would like to point out that the approach of transforming the convex constraint into a linear operator with range in the space of continuous functions with compact support was first used in [35]. Later, it was used in [21] in conjunction with Theorem 4.2.2 to prove no duality gap for problems with infinite-size infinite dimensional LMI's constraints. The results of [21] however, follow more naturally and directly if one uses the representation of the LMI constraints given in Section 3.3.4.

6.2.1 Example

We present now a simple example of ℓ_1 sensitivity minimization with a frequency point magnitude constraint. We will see that the problem presented is actually equivalent to a finite dimensional convex optimization. Consider the following plant P, with λ-transform

$$\hat{P}(\lambda) = \frac{1-2\lambda}{1-.5\lambda}.$$

We desire to minimize the ℓ_1 norm of the sensitivity function $S = (1-PK)^{-1}$ and to attenuate, by a factor of ten, a sinusoidal input at angular frequency $\omega = \pi/6$. The problem can be formally posed as follows:

$$\mu^o = \inf_{K-stab} \|(1-PK)^{-1}\|_1 \tag{6.9}$$

subject to:

$$|\hat{S}(\pi/6)| \leq 0.1$$

There is only one feasibility constraint for this problem, given by the presence of a zero of $\hat{P}(\lambda)$ inside the unit disc, precisely at $\lambda = 0.5$. To achieve internal stability, the controller cannot cancel such zero, thus, all the achievable (stable) sensitivity functions must satisfy $\hat{S}(0.5) = 1$. Problem (6.9) can be rewritten as follows:

$$\mu^o = \inf \|\Phi\|_1 \tag{6.10}$$

subject to:

$$A_{feas}\Phi = 1$$
$$|\hat{\Phi}(\pi/6)| \leq 0.1$$

with $A_{feas} = [1 \;\; \frac{1}{2} \;\; \frac{1}{4} \;\; \cdots \;\; \frac{1}{2^n} \;\; \ldots]$.

Without considering the magnitude constraint, the optimal ℓ_1 solution would be $\hat{\Phi} = 1$, constant for all ω. Since we are requiring $|\hat{\Phi}(\pi/6)| \leq 0.1$, the magnitude constraint is going to be active in the optimization. The dual of Problem (6.10) is given by Problem (6.8) with $\omega = \pi/6$.

There is an infinite number of constraints in the dual problem, given by

$$|(A^*_{feas})_n z^*_1 + \alpha\cos(n\omega+\theta)| \leq 1 \quad \text{for } n = 0, 1, \ldots \tag{6.11}$$

Notice that there is an implicit constraint in the dual problem. Given that the only row of A_{feas} is a sequence in c_0, as n goes to infinity, the term $(A^*_{feas})_n z^*_1$ tends to zero in (6.11). Thus, in the limit, the following constraint on α and θ is imposed:

$$|\alpha\cos(n\omega+\theta)| \leq 1. \tag{6.12}$$

If ω is an irrational multiple of π, the implicit constraint is given by $|\alpha| \leq 1$, while if ω is a rational multiple of π, say $\omega = \frac{p}{q}\pi$, then $\cos(n\frac{p}{q}\pi+\theta)$ is periodic after $2*q$ samples for a fixed θ.

Having noticed that, we now turn our attention to feasible suboptimal FIR solutions. In particular we derive a sufficient condition that guarantees that the optimal primal solution is indeed *FIR*. We can obtain a feasible suboptimal FIR solution for the primal problem by imposing that $\Phi(k) = 0$ for all $k \geq N$, for some N. The dual of this problem is given by Problem (6.8) where only the first N constraints are present. Thus the dual solution, based only on the first N dual constraints, may or may not satisfy the implicit constraint (6.12). If it does, then, the implicit constraint is inactive and therefore there exists some $M \geq N$ such that all the constraints after the Mth will be inactive in the optimization. Thus, from the alignment condition, the optimal primal solution has finite support.

For the example we are considering the implicit constraint is inactive for $N \geq 7$. The optimal dual solution for $N = 7$ is given by $z^*_{1,7} = 1.9692$, $\alpha_7 = -0.9692$, and $\theta_7 = 0$. The optimal cost is $\mu^o_7 = 1.8723$. The solution does not seem to change for $N \geq 7$. In particular, it is the same for $N = 13$. To see that the above is indeed the optimal dual solution, we use the periodicity of the term involving $\cos(n\frac{\pi}{6} + \theta)$. This term is periodic after the first 12 samples. Since $(A_{feas}z^*_{1,13})_n \leq \frac{1}{2^{13}}1.9692 = 2.403810^{-4}$, for all $n \geq 13$, then, $|A_{feas}z^*_{1,13} + \alpha_{13}\cos(n\frac{\pi}{6} + \theta_{13})| < 1$ for all $n \geq 13$. Thus, the only active constraints are at most the first thirteen, and hence the problem is indeed finite dimensional. The resulting optimal FIR primal solution is of order 6 and it is given by:

$$\hat{\Phi}(\lambda) = 0.9862 + 0.8802\lambda^6.$$

The optimal controller K, which turns out to be of order 7, can be computed from the expression: $\Phi = (1 - PK)^{-1}$.

6.3 Linear Programming Approximation

In this section, we want to investigate the approximation of Problem (6.7) and its dual, Problem (6.8), by finite dimensional linear programming problems (the final objective is to derive finite dimensional linear programming problems that approximate Problem (6.7) and Problem (6.8)). Even in the SISO or in the one-block case, Problem (6.7) has an infinite number of variables and an infinite number of linear constraints. As already mentioned the infinite number of linear constraints is needed to represent the one convex magnitude constraint. As a first step we study the approximation of the convex constraint by only a finite number of linear constraints.

6.3.1 Approximation of the Magnitude Constraint with a Finite Number of Linear Constraints

We want to investigate what happens if the continuum of constraints represented by $\mathcal{A}_{\mathcal{H}_\infty}\Phi \leq \gamma$ in Problem (6.7) is replaced with only a finite number of linear constraints. The typical method used in continuous LP is to sample the continuous variable in a finite number of points. Next theorem shows that the finer the sampling is, the closer the approximation is to the original problem.

Theorem 6.3.1. *Let θ_i $i = [1, \ldots, k]$ represent equally spaced samples of the unit circle so that the k arcs into which the circle is partitioned are of equal length $\Delta_i = 2\pi/k < \pi$. Assume that $\gamma > 0$ and consider the following linear* *program, whose cost is a function of the value of the magnitude constraint* $y \geq 0$*:*

$$\mu(y) = \inf_{\Phi \in \ell_1^{n_z \times n_w}} \|\Phi\|_1 \tag{6.13}$$

subject to:

$$\mathcal{A}_{feas}\Phi = b_{feas}$$
$$\mathcal{A}^k_{\mathcal{H}_\infty}\Phi \leq 1y$$

where $1y$ *is a column vector of length* k *whose elements are all equal to* y*, and where* $\mathcal{A}^k_{\mathcal{H}_\infty} : \ell_1 \to \mathbb{R}^k$ *acts on* $\Phi \in \ell_1^{n_z \times n_w}$ *as follows:*

$$(\mathcal{A}^k_{\mathcal{H}_\infty}\Phi)_l = \sum_{n=0}^{\infty} \Phi_{11}(n)\cos(n\omega + \theta_l), \quad l = 1, \dots, k,$$

Then,

$$\mu(\gamma) \leq \mu^o \leq \mu(\gamma\cos(\Delta/2))$$

Moreover,

$$\mu(\gamma\cos(\Delta/2)) - \mu^o \leq \gamma(\cos(\Delta/2) - 1)\langle 1, \rho\rangle,$$

where $\langle 1, \rho\rangle = \sum_{l=1}^k \rho_l$ *and the* ρ_l*'s* $l = 1, \dots, k$ *are optimal dual variables for the following problem:*

$$\max_{z_1^*, \rho} \quad \langle b_{feas}, z_1^*\rangle + \gamma\cos(\Delta/2)\langle 1, \rho\rangle$$

subject to:

$$\|\mathcal{A}^*_{feas} z_1^* + \mathcal{A}^{k*}_{\mathcal{H}_\infty}\rho\|_\infty \leq 1 \tag{6.14}$$
$$\rho \leq 0,\ \rho \in \mathbb{R}^k,\ z_1^* \in \ell_\infty$$

which is the dual, with no duality gap, of Problem (6.13) with $y = \gamma\cos(\Delta/2)$.

Proof. Clearly the optimal cost, μ^o, of problem (6.7) is always greater of $\mu(\gamma)$. It is left to the reader to verify that the dual of Problem (6.13) with $y = \gamma\cos(\Delta/2)$ is indeed given by Problem (6.14), where $\mathcal{A}^{k*}_{\mathcal{H}_\infty}$ acts on $\rho \in \mathbb{R}^k$ as follows:

$$[\mathcal{A}^{k*}_{\mathcal{H}_\infty}\rho]_{ij} = \begin{cases} \sum_{l=1}^k \rho_l\cos(\theta_l)\cos(n\omega) - \sum_{l=1}^k \rho_l\sin(\theta_l)\sin(n\omega), & i = j = 1 \\ 0, & \begin{array}{l} i = 2, \dots, n_z \\ j = 2, \dots, n_w \end{array} \end{cases}$$

The assumption that γ is strictly greater than 0 guarantees that the regularity condition in Theorem 4.2.3 is satisfied. See Remark 6.2.1.

First thing we are going to show is that the optimal solution for problem (6.14) will have at most two of the ρ_i's strictly less than zero, and all the others will be equal to 0. This is equivalent to say that at most two of the k constraints of $\mathcal{A}^k_{\mathcal{H}_\infty}\Phi \leq \gamma\cos(\Delta/2)$ are going to be active in the optimization. Consider, without loss of generality, the case with $k = 3$. Let $\alpha < 0$ and θ be such that

$$\alpha\cos(\theta) = \rho_1\cos(\theta_1) + \rho_2\cos(\theta_2) + \rho_3\cos(\theta_3)$$
$$\alpha\sin(\theta) = \rho_1\sin(\theta_1) + \rho_2\sin(\theta_2) + \rho_3\sin(\theta_3)$$

where ρ_1, ρ_2, ρ_3 are optimal for Problem (6.14). Notice that the above equations always have a solution with $\alpha \leq 0$ and $\theta \in [0, 2\pi)$. Suppose without loss of generality that $\theta \in [\theta_2, \theta_1]$. It is easy to derive the following expression:

$$\rho_1 + \rho_2 + \rho_3 = \alpha\frac{\cos(\overline{\theta} - \theta)}{\cos(\Delta/2)} + \rho_3\left(1 - \frac{\cos(\overline{\theta} - \theta_3)}{\cos(\Delta/2)}\right) \tag{6.15}$$

where $\overline{\theta} = (\theta_2 + \theta_1)/2$. Since for $\theta_3 = \overline{\theta} + (3/2)\Delta$, the term multiplying ρ_3 in the right hand side of Equation (6.15) is positive, it follows that

$$\rho_1 + \rho_2 + \rho_3 \leq \alpha\frac{\cos(\overline{\theta} - \theta)}{\cos(\Delta/2)} \tag{6.16}$$

This implies that, in this case, the optimal solution will have $\rho_3 = 0$. Actually, the relation above holds for any $\theta_3 \notin [\theta_2, \theta_1]$ in Equation (6.15). This last observation allows us to deduce that for any k the optimal solution for (6.14) will have at most two non-zero ρ_l's. Moreover, if we call such non-zero elements ρ_s and ρ_p with $p > s$, then $p = (s+1) \mod k$.
Finally suppose that $\alpha < 0$ and θ are the optimal solution for (6.8), then given the partition $[\theta_1, \ldots, \theta_k]$, θ will fall at least into one of the k intervals. Let $\theta \in [\theta_{i+1}, \theta_i]$. Reparametrizing the optimal solution as follows:

$$\begin{aligned}\alpha\cos(\theta) &= \rho_i\cos(\theta_i) + \rho_{i+1}\cos(\theta_{i+1})\\ \alpha\sin(\theta) &= \rho_i\sin(\theta_i) + \rho_{i+1}\sin(\theta_{i+1})\end{aligned} \tag{6.17}$$

we obtain that: $\rho_i + \rho_{i+1} = \alpha\dfrac{\cos(\overline{\theta} - \theta)}{\cos(\Delta/2)}$. Since θ is arbitrary in $[\theta_{i+1}, \theta_i]$, the following inequalities hold:

$$(\rho_i + \rho_{i+1}) \leq \alpha \leq (\rho_i + \rho_{i+1})\cos(\Delta/2)$$

Thus, any optimal solution for (6.8), re-parametrized according to Equations (6.17), is a suboptimal solution for the dual Problem in (6.14) and therefore $\mu^o \leq \mu(\gamma\cos(\Delta/2))$.

Since the optimal solution for the dual of $\mu(\gamma\cos(\Delta/2))$ is a feasible solution for the dual of $\mu(\gamma)$, then the error of solving $\mu(\gamma\cos(\Delta/2))$ instead of problem (6.7) is less than

$$\gamma(\cos(\Delta/2) - 1)\langle 1, \rho\rangle$$

■

Thus, problem (6.7) can be approximated arbitrarily well by a linear program as Δ goes to 0. Note however that Problem (6.14) is still infinite dimensional.

6.3.2 Finite Dimensional Approximation

There are two factors that make Problem (6.14) infinite dimensional.

(i) In the one-block case there are no rank interpolation conditions, thus $z_1^* \in \mathbb{R}^{c_z}$ for some finite c_z. Although in this case the range of $\mathcal{A}_{feas}^*$ is in $c_0^{n_z \times n_w}$ and contrary to what happens in the standard ℓ_1 one-block problem, there may not exist a finite N after which the dual constraints are all inactive. This is due to the presence of the sine and cosine terms that are not elements of c_0 and make this problem similar to the standard ℓ_1 problem with interpolations on the unit circle. Although we will not present the proof here, see [1], it is possible to obtain a finite dimensional approximation to this semi-infinite LP within any pre-defined tolerance.

(ii) In the multi-block case, the optimal z_1^* is an element in ℓ_∞. Although there will be no duality gap, it is unclear how to compute exact or approximate solutions for such problem. We will show shortly that sequences of dual approximations with finite support of increasing length may fail to converge to the optimal cost. This is due to the fact that finite support sequences are not dense in ℓ_∞.

To keep the notation simple we drop the superscript k from $\mathcal{A}_{\mathcal{H}_\infty}^k$, without loss of generality we may assume that $\mathcal{A}_{\mathcal{H}_\infty}$ maps $\ell_1^{n_z \times n_w}$ to $\mathbb{R}^k$. We also assume that $\gamma > 0$. The primal problem we consider for the rest of the section is:

$$\mu^o = \inf_{\Phi \in \ell_1^{n_z \times n_w}} \|\Phi\|_1 \tag{6.18}$$

subject to:

$$\mathcal{A}_{feas}\Phi = b_{feas},$$
$$\mathcal{A}_{\mathcal{H}_\infty}\Phi \le \gamma$$

For future reference, we write the dual of Problem (6.18).

$$\max_{z_1^*, z_2^*} \langle b_{feas}, z_1^* \rangle + \langle \gamma 1, z_2^* \rangle \tag{6.19}$$

subject to:

$$\|\mathcal{A}_{feas}^* z_1^* + \mathcal{A}_{\mathcal{H}_\infty}^* z_2^*\|_\infty \le 1$$
$$z_2^* \le 0$$
$$z_1^* \in \ell_\infty,\ z_2^* \in \mathbb{R}^k$$

From Theorem 4.2.3 we know that there is no duality gap between Problem (6.18) and Problem (6.19).

Before we begin the study of finite dimensional approximations of the dual problem in (6.19), we briefly consider finite dimensional approximation of the primal problem in (6.18). In particular we want to understand if the cost of finite support feasible primal solutions can approximate, to any accuracy, the optimal cost μ^o for Problem(6.18). We must point out that Problem(6.18) may not have any finite support feasible solution. However in the case a feasible finite support solution exists, it is possible to approximate μ^0 from above arbitrarily well by a finite support feasible sequence of adequate length. The proof is similar to the proof of Theorem 5.3.2 and therefore it is omitted.

We now want to investigate whether the optimal cost of Problem (6.19) can be achieved by constraining z_1^*, in Problem (6.19), to be in c_0 instead of ℓ_∞. Formally, such problem is given by:

$$\nu^o = \sup_{z_1^*, z_2^*} \langle b_{feas}, z_1^* \rangle + \langle \gamma 1, z_2^* \rangle \tag{6.20}$$

subject to:

$$\left\| \mathcal{A}_{feas}^* z_1^* + \mathcal{A}_{\mathcal{H}_\infty}^* z_2^* \right\|_\infty \leq 1$$

$$z_2^* \leq 0,$$

$$z_1^* \in c_0, \ z_2^* \in \mathbb{R}^k$$

Notice that, if $\mu^0 > \nu^o$, then no finite support feasible solution of Problem (6.19) can approximate arbitrarily close the optimal dual cost μ^o.

In what follows we characterize Problem (6.20) as the (pre-)dual, with no duality gap, of a particular minimization problem whose optimal cost may differ from μ^o in the multi-block case. Notice that Problem (6.20) is not the (pre-)dual of Problem (6.18). The operator $\mathcal{A} = \begin{bmatrix} \mathcal{A}_{feas} \\ \mathcal{A}_{\mathcal{H}_\infty} \end{bmatrix} : c_0^* \to c_0^* \times \mathbb{R}^k$ is not the adjoint of an operator from $c_0 \times \mathbb{R}^k$ to c_0.

We assume, without loss of generality, that the magnitude constraint at one frequency point is imposed on Φ_{11}. Let us define the following vector space: $X = c_0^{n_z \times n_w} \oplus s_1 \oplus s_2$, where

$$s_1 = \left\{ y \in \ell_\infty^{n_z \times n_w} \mid \begin{cases} y_{11}(n) = \alpha_1 \cos(n\omega_0) \text{ with } \alpha_1 \in \mathbb{R} \\ y_{ij}(n) = \quad 0 \quad \forall\, n \quad \text{if } i \neq 1, j \neq 1 \end{cases} \right\}$$

$$s_2 = \left\{ y \in \ell_\infty^{n_z \times n_w} \mid \begin{cases} y_{11}(n) = \alpha_2 \sin(n\omega_0) \text{ with } \alpha_2 \in \mathbb{R} \\ y_{ij}(n) = \quad 0 \quad \forall\, n \quad \text{if } i \neq 1, j \neq 1 \end{cases} \right\}$$

and $\oplus$ indicates direct sum. Equipped with the ℓ_∞ norm, X is a closed subspace of $\ell_\infty^{n_z \times n_w}$ and hence a Banach space.

We need now to characterize X^*. Toward this end, consider the following functions on X.

$$f_1 : X \to \mathbb{R}, \ f_1(x) = \lim_{N \to \infty} \frac{2}{N} \sum_{i=1}^{N} x_{11}(i) \cos(i\omega_0)$$

$$f_2 : X \to \mathbb{R}, \ f_2(x) = \lim_{N \to \infty} \frac{2}{N} \sum_{i=1}^{N} x_{11}(i) \sin(i\omega_0)$$

It is left to the reader to verify that, for any element $x \in X$, the $f_i(\cdot)$ for $i = 1, 2$ are well defined and linear and, for any $x \in c_0^{n_z \times n_w}$, $f_i(x) = 0$. Also notice that, if $x \in s_1$, then, $f_1(x) = \alpha_1$ and $f_2(x) = 0$, while, if $x \in s_2$, then, $f_1(x) = 0$, and $f_2(x) = \alpha_2$. Therefore, for any $w \in \ell_1^{n_z \times n_w}$, $\beta_1, \beta_2 \in \mathbb{R}$,

$$x^* = w + \beta_1 f_1(\cdot) + \beta_2 f_2(\cdot)$$

is a linear functional on X.

We claim that any linear functional on X can be represented in the above form. To prove it, it is sufficient to show that, if $\langle x, x^* \rangle = 0$ for all the linear functional x^* with the above representation, then $x = 0$. This is true, since $f_1(x) = 0$ implies that $\alpha_1 = 0$ and $f_2(x) = 0$ implies that $\alpha_2 = 0$. Hence x must be an element in $c_0^{n_z \times n_w}$, which must be the zero element, since $\ell_1^{n_z \times n_w} = (c_0^{n_z \times n_w})^*$.

It is not difficult to verify that the induced norm of x^* is given by

$$\|x^*\| = \|w\|_1 + \|[\beta_1, \beta_2]\|$$

where

$$\|[\beta_1, \beta_2]\| = \sup_{\|s\|_\infty \leq 1} |\beta_1 f_1(s) + \beta_2 f_2(s)| \qquad \text{with } s \in s_1 \oplus s_2.$$

For any $\omega \in [0, 2\pi)$, the value of $\|[\beta_1, \beta_2]\|$ can be computed by solving the following linear programming problem:

$$\begin{aligned} \|[\beta_1, \beta_2]\| = & \max_{\alpha_1, \alpha_2} \quad \beta_1 \alpha_1 + \beta_2 \alpha_2 \\ & \text{subject to:} \\ & |\alpha_1 \cos(n\omega) + \alpha_2 \sin(n\omega)| \leq 1, \quad n = 0, 1, 2, \ldots \\ & \alpha_1, \alpha_2 \in \mathbb{R} \end{aligned} \tag{6.21}$$

If ω is an irrational multiple of π, then, the constraints on α_1 and α_2 are equivalent to the constraint: $\|[\alpha_1, \alpha_2]\|_2 \leq 1$. In this case,

$$\|[\beta_1, \beta_2]\| = \|[\beta_1, \beta_2]\|_2.$$

If ω is a rational multiple of π, say $\omega = p/q 2\pi$, then, the sequences $\cos(n\omega)$ and $\sin(n\omega)$ are periodic with period at most q. This implies that the constraints in (6.21) keep repeating after the first q. Therefore, we can retain the first q constraints and omit the rest from the optimization without changing the value of the optimal cost. Thus, the value of $\|[\beta_1, \beta_2]\|$ can be computed by solving a finite dimensional linear programming problem.

We can associate a unique element $y^* = [w, \beta_1, \beta_2] \in Y^* = \ell_1^{n_z \times n_w} \times \mathbb{R}^2$ to any element $x^* \in X^*$ and vice versa. If the norm of y^* is defined to be $\|y^*\| = \|w\|_1 + \|[\beta_1, \beta_2]\|$, then the mapping from X^* to $\ell_1^{n_z \times n_w} \times \mathbb{R}^2$ is an isometric isomorphism. See [23] for the definition. Y^* allows us to obtain a concrete representation for the following optimization problem in X^*.

We consider the following optimization problem in the space X^*:

$$\begin{aligned} \nu^o = & \inf_{x^* \in X^*} \quad \|x^*\| \\ & \text{subject to:} \\ & \bar{A}_{feas} x^* = b_{feas}, \\ & \bar{A}_{\mathcal{H}_\infty} x^* \leq 1\gamma \end{aligned} \tag{6.22}$$

where $\bar{A}_{\mathcal{H}_\infty} : X^* \to \mathbb{R}^k$ and $\bar{A}_{feas} : X^* \to \ell_1 = c_0^*$ are defined as follows: for each $x^* \in X^*$, let $y^* = [\Phi, \beta]$ be the unique respective element in $Y^* = \ell_1^{n_z \times n_w} \times \mathbb{R}^2$. Then,

$$\bar{\mathcal{A}}_{\mathcal{H}_\infty} x^* = \mathcal{A}_{\mathcal{H}_\infty} \Phi + \mathcal{A}_0 \beta$$

with

$$\mathcal{A}_0 = \begin{bmatrix} \cos(\theta_1) & -\sin(\theta_1) \\ \vdots & \vdots \\ \cos(\theta_k) & -\sin(\theta_k) \end{bmatrix}, \tag{6.23}$$

$$\bar{\mathcal{A}}_{feas} x^* = \mathcal{A}_{feas} \Phi. \tag{6.24}$$

We are now going to show that the operator $\bar{\mathcal{A}} = \begin{bmatrix} \bar{\mathcal{A}}_{feas} \\ \bar{\mathcal{A}}_{\mathcal{H}_\infty} \end{bmatrix}$ is the adjoint of an operator, ${}^*\bar{\mathcal{A}}$, from $c_0 \times \mathbb{R}^2$ to X.

Lemma 6.3.1.

(i) Let ${}^*\mathcal{A}_{feas} : c_0 \to c_0^{n_z \times n_w}$ be the pre-adjoint of $\mathcal{A}_{feas} : \ell_1^{n_z \times n_w} \to \ell_1$ which is equal to $\mathcal{A}_{feas}^*$ with domain restricted to c_0. Then, ${}^*\bar{\mathcal{A}}_{feas} : c_0 \times \mathbb{R}^2 \to X$ is given by ${}^*\mathcal{A}_{feas}$ ([15]).

(ii) Let $\mathcal{A}_{\mathcal{H}_\infty}^* : \mathbb{R}^k \to \ell_\infty^{n_z \times n_w}$ be the adjoint of $\mathcal{A}_{\mathcal{H}_\infty} : \ell_1^{n_z \times n_w} \to \mathbb{R}^k$. Then, ${}^*\bar{\mathcal{A}}_{\mathcal{H}_\infty} : c_0 \times \mathbb{R}^2 \to X$, the pre-adjoint of $\bar{\mathcal{A}}_{\mathcal{H}_\infty}$, is given by $\mathcal{A}_{\mathcal{H}_\infty}^*$.

(iii) Finally, ${}^*\bar{\mathcal{A}} = [\, \mathcal{A}_{feas}^*, \;\; \mathcal{A}_{\mathcal{H}_\infty}^* \,]$ with domain restricted to $c_0 \times \mathbb{R}^2$.

Proof.

(i) Let $x = \mathcal{A}_{feas}^* z_1$. For any $z_1 \in c_0$, x belongs to $c_0^{n_z \times n_w} \subset X$. Thus for any $x^* \in X^*$ we have that

$$\langle x, x^* \rangle = \beta_1 f_1(x) + \beta_2 f_2(x) + \sum_{i=1}^{n_z} \sum_{j=1}^{n_w} \sum_{t=0}^{\infty} w_{ij}(t) x_{ij}(t)$$

Since $x \in c_0^{n_z \times n_w}$, $f_1(x) = f_2(x) = 0$. Note also that, for $w \in \ell_1^{n_z \times n_w}$, the last term of the above expression is nothing but the standard linear functional on $c_0^{n_z \times n_w}$. Thus $\langle x, x^* \rangle = \langle x, w \rangle$ and $w \in \ell_1^{n_z \times n_w}$. Substituting the expression for x, we have that:

$$\langle \mathcal{A}_{feas}^* z_1, x^* \rangle = \langle \mathcal{A}_{feas}^* z_1, w \rangle = \langle z_1, \mathcal{A}_{feas} w \rangle = \langle z_1, \bar{\mathcal{A}}_{feas} x^* \rangle$$

(ii) First notice that actually the range of $\mathcal{A}_{\mathcal{H}_\infty}^*$ is in X. Let $x = \mathcal{A}_{\mathcal{H}_\infty}^* z_2$ with $z_2 \in \mathbb{R}^k$. Also notice that, given the structure of $\mathcal{A}_{\mathcal{H}_\infty}$, $x \in s_1 \oplus s_2$. Then,

$$\langle x, x^* \rangle = \langle \mathcal{A}_{\mathcal{H}_\infty}^* z_2, x^* \rangle = \beta_1 f_1(x) + \beta_2 f_2(x) + \sum_{i=1}^{n_z} \sum_{j=1}^{n_w} \sum_{t=0}^{\infty} w_{ij}(t) x_{ij}(t)$$

Since, by hypothesis, only Φ_{11} is affected by the extra constraints, the last term in the above equation reduces to $\sum_{t=0}^{\infty} w_{11}(t) x_{11}(t)$. Substituting the expression for x we have that

$$\begin{aligned}\langle \mathcal{A}^*_{\mathcal{H}_\infty} z_2, x^*\rangle &= \beta_1 \textstyle\sum_{l=1}^k \cos(\theta_l) z_{2l} - \beta_2 \sum_{l=1}^k \sin(\theta_l) z_{2l} + \\ &+ \sum_{t=0}^{\infty} w_{11}(t) \left(\sum_{l=1}^{k} \cos(\theta_l) z_{2l}\right) \cos(t\omega_0) + \\ &- \sum_{t=0}^{\infty} w_{11}(t) \left(\sum_{l=1}^{k} \sin(\theta_l) z_{2l}\right) \sin(t\omega_0) \\ &= z_2^T \left[\mathcal{A}_0 \beta + \mathcal{A}_{\mathcal{H}_\infty} w\right] \\ &= \langle z_2, \bar{\mathcal{A}}_{\mathcal{H}_\infty} x^*\rangle\end{aligned}$$

(iii) This follows immediately from the property of the adjoint operator. ∎

Given the result above, we are now able to write the (pre-)dual of Problem (6.22) in $c_0 \times \mathbb{R}^2$ as follows:

$$\begin{aligned}&\sup_{z_1, z_2} \quad \langle b_{feas}, z_1\rangle + \langle \gamma 1, z_2\rangle, \\ &\text{subject to:} \\ &\quad \left\|\mathcal{A}^*_{feas} z_1 + \mathcal{A}^*_{\mathcal{H}_\infty} z_2\right\|_\infty \leq 1, \\ &\quad z_2 \leq 0, \\ &\quad z_1 \in c_0,\ z_2 \in \mathbb{R}^k.\end{aligned} \tag{6.25}$$

Notice that this problem is nothing but Problem (6.20). Problem (6.22) can be concretely represented in the space Y^*. The resulting equivalent problem is given by:

$$\begin{aligned}\nu^o = &\inf_{\Phi, \beta} \quad \|\beta\| + \|\Phi\|_1, \\ &\text{subject to:} \\ &\quad \mathcal{A}_{feas}\Phi = b_{feas}, \\ &\quad \mathcal{A}_0 \beta + \mathcal{A}_{\mathcal{H}_\infty} \Phi \leq 1\gamma, \\ &\quad \beta \in \mathbb{R}^2, \Phi \in \ell_1^{n_z \times n_w}.\end{aligned} \tag{6.26}$$

It is easy to see that Problem (6.26) is a restatement in Y^* of Problem (6.22). Each row of $\mathcal{A}_{feas}$ belongs to $c_0^{n_z \times n_w}$, therefore x^* acts on it as an element in $\ell_1^{n_z \times n_w}$. The coefficients of β_1 and β_2 are zero since $f_j((\mathcal{A}_{feas})_i) = 0$ for $j = 1, 2$. The action of x^* on rows of $\mathcal{A}_{\mathcal{H}_\infty}$ produces the following:

$$\langle (\mathcal{A}_{\mathcal{H}_\infty})_i, x^*\rangle = \beta_1 \cos(\theta_i) - \beta_2 \sin(\theta_i) + (\mathcal{A}_{\mathcal{H}_\infty})_i \Phi$$

Next theorem states that there is no gap between Problem (6.26) and Problem (6.25). Notice that Problem (6.26) is feasible for all $\gamma \geq 0$.

Theorem 6.3.2. *Consider Problem (6.26). Assume that $\gamma > 0$. Then, the optimal cost of Problem (6.26) is equal to the optimal cost of Problem (6.25). Moreover, an optimal minimizing solution for Problem (6.26) always exists.*

Proof. Apply Theorem 4.2.3 to Problem (6.25) rewritten as follows:

$$
\begin{aligned}
-\inf \quad & \langle -b_{feas}, z_1\rangle + \langle -\gamma 1, z_2\rangle, \\
\text{subject to:} \quad & w +^* \bar{A}_{feas} z_1 +^* \bar{A}_{\mathcal{H}_\infty} z_2 = 0, \\
& \eta = 1, \\
& \|w\|_\infty \leq \eta,\ z_2 \leq 0, \\
& w \in X,\ z_1 \in c_0,\ z_2 \in \mathbb{R}^k,\ \eta \in \mathbb{R}^+.
\end{aligned}
$$

The details are left to the reader. This establishes no duality gap between Problem (6.25) and Problem (6.22). The result follows from the equivalence of Problem (6.22) with Problem (6.26). ■

Summarizing the development so far, we have that any finite support feasible solution of Problem (6.19) will have an optimal cost always smaller than ν^o. Moreover, the optimal cost ν^o can be approximated arbitrarily well by a finite support feasible solution of adequate length.

With the above theorem, we have obtained a characterization without gap of the primal problem (Problem (6.26)), whose (pre-)dual optimal cost can be approximated arbitrarily well by finite support feasible solutions. Next Theorem investigates the relationship with the original problem.

Theorem 6.3.3. *Consider Problem (6.26) and Problem (6.18). Then,*

$$\nu^o \leq \mu^o.$$

If the problem is one-block, then $\nu^o = \mu^o$.

Proof. To any feasible solution Φ of Problem (6.18) corresponds a feasible solution $[\Phi, 0]$ of Problem (6.26) with the same cost, but not vice versa. Thus, it must be $\nu^o \leq \mu^o$.

If the problem is one-block, then $\mathcal{A}^*_{feas}$ maps $\mathbb{R}^{c_z}$ to $c_o^{n_z \times n_w}$. In this case, Problem (6.20) and Problem (6.19) are indeed the same problem. Since both have no duality gap with their respective primal problems, it follows that $\nu^o = \mu^o$. ■

For multi-block problems the inequality may be strict. To show that for multi-block problems it may happen that $\nu^o < \mu^o$ we present a simple two-block row example.

6.3.3 Counterexample for Multi-Block Problems

Consider the following set of closed loop stable maps:

$$(\phi_1, \phi_2) = (h_1, h_2) + q(v_1, v_2) \tag{6.27}$$

where $q \in \ell_1$, $\hat{h}_1(\lambda) = 0$, $\hat{h}_2(\lambda) = -10$, $\hat{v}_1(\lambda) = 1/3$, and $\hat{v}_2(\lambda) = 1 + 0.5\lambda$. Since $\hat{V} = (\hat{v}_1, \hat{v}_2)$ has no right zeros in the unit disc, there are no interpolation conditions. The rank interpolation conditions are given by the following constraints:

$$(\hat{\phi}_1, \hat{\phi}_2)\begin{pmatrix} \hat{v}_2 \\ -\hat{v}_1 \end{pmatrix} = (\hat{h}_1, \hat{h}_2)\begin{pmatrix} \hat{v}_2 \\ -\hat{v}_1 \end{pmatrix}$$

Substituting the value for $\hat{h}_1$, $\hat{h}_2$ and $\hat{v}_1$, we have:

$$\hat{\phi}_1(\lambda)\hat{v}_2(\lambda) - (1/3)\hat{\phi}_2(\lambda) = 10/3$$

or equivalently:

$$(\hat{v}_2(\lambda), -1/3)\begin{pmatrix} \hat{\phi}_1(\lambda) \\ \hat{\phi}_2(\lambda) \end{pmatrix} = 10/3$$

The above constraints transformed in the space of sequences are represented as

$$[V_2, -(1/3)I]\begin{bmatrix} \phi_1 \\ \phi_2 \end{bmatrix} = b$$

where V_2 is the Toeplitz infinite matrix representing the convolution with $\hat{v}_2(\lambda)$ and $b = \{10/3, \ 0, \ldots\}$.

Optimal cost of Problem (6.18). We want to solve the following constrained optimization problem:

$$\mu^o = \inf \quad \|[\phi_1, \phi_2]\|_1, \tag{6.28}$$

subject to:

$$[V_2, -\tfrac{1}{3}I]\begin{bmatrix} \phi_1 \\ \phi_2 \end{bmatrix} = b,$$

$$\sum_{t=0}^{\infty} \cos(t\omega_0)\phi_1(t) \leq 1.$$

Here,

$$\begin{aligned} \mathcal{A}_{\mathcal{H}_\infty} &= [A_{1,\mathcal{H}_\infty}, \ 0] \\ &= [[1, \cos(\omega_0), \cos(2\omega_0), \ldots, \cos(t\omega_0), \ldots], \ [0, 0, \ldots]] \end{aligned}$$

Note that, this is a special case of Problem (6.18) or Problem (6.13) with $k = 1$, i.e., only one sample of the unit circle is considered. In particular, we have chosen the point $\theta = 0$ at a frequency $\omega_0 = \pi/6$.

We compute the exact value for μ^o as described next. We provide a feasible, and hence sub-optimal, solution to problem (6.28), and a feasible solution to the dual of Problem (6.28). Then, we show that both solutions have the same cost and therefore they must be the optimal primal and dual solutions.

Pick $q = A/(1 + 0.5\lambda)$ in Equation (6.27), where the positive constant A has to be determined. For any $A \geq 0$, the norm of Φ is given by:

$$\|\Phi\|_1 = \|\Phi_1\|_1 + \|\Phi_2\|_1 = \|\tfrac{A}{3}\tfrac{1}{1+0.5\lambda}\|_1 + |A - 10| = (2A)/3 + |A - 10|$$

We select A so that Φ_1 satisfies the constraint:

$$\sum_{t=0}^{\infty} \cos(t\omega_0)\phi_1(t) \leq 1$$

Substituting the expression for Φ_1, we obtain that:

$$(A/3)\sum_{t=0}^{\infty} \cos(t\omega_0)(-1/2)^t = \frac{A}{3}\frac{1+0.5\cos(\omega_0)}{1.25+\cos(\omega_0)}.$$

Thus, the constraint is satisfied for all A such that:

$$A \leq 3\frac{1.25+\cos(\omega_0)}{1+0.5\cos(\omega_0)} = 4.4299$$

For $A = 4.4299$, the relative solution is feasible for Problem (6.28) and it results that $\|\Phi\|_1 = 8.5234$.

The dual of Problem (6.28) is given by the following optimization in $\ell_\infty \times \mathbb{R}$:

$$\mu^o = \max_{z_1,z_2} \quad \tfrac{10}{3}z_{11} + z_2, \tag{6.29}$$

subject to:

$$\|V_2^* z_1 + A_{1,\mathcal{H}_\infty}^T z_2\|_\infty \leq 1,$$
$$\|\tfrac{1}{3}Iz_1\|_\infty \leq 1,$$
$$z_2 \leq 0,\ z_1 \in \ell_\infty,$$

where z_{11} represents the first element of the sequence $z_1 \in \ell_\infty$.

A feasible solution for the dual problem can be computed as follows. For any $z_2 \leq 0$, let z_1 be sum of two sequences, $z_1 = w_1 + w_2$, with

$$w_1 = -(V_2^*)^{-1} z_2 \cos(t\omega_0)$$

and w_2 such that

$$\|V_2^* w_2\|_\infty \leq 1.$$

Clearly,

$$\|V_2^* z_1 + z_2\cos(t\omega_0)\|_\infty \leq 1$$

is satisfied for such a choice of z_2 and z_1. Among the possible w_2, let's select the following alternating sequence: $w_2 = \{2, -2, 2, -2, 2, \ldots\}$. It is easy to see that $\|V_2^* w_2\|_\infty \leq 1$.

It is left to reader to verify that:

$$\begin{aligned} w_1(n) &= -z_2\left(\cos(n\omega_0)\sum_{t=0}^{\infty}\cos(t\omega_0)(-\frac{1}{2})^t - \sin(n\omega_0)\sum_{t=0}^{\infty}\sin(t\omega_0)(-\frac{1}{2})^t\right) \\ &= -z_2\left(\frac{1+0.5\cos(\omega_0)}{1.25+\cos(\omega_0)}\cos(n\omega_0) - \frac{0.5\sin(\omega_0)}{1.25+\cos(\omega_0)}\sin(n\omega_0)\right) \end{aligned} \tag{6.30}$$

For the choice of ω_0, w_1 is a periodic sequence with 12 samples in each period. By computing the value of $w_1(n)$ for the first 12 samples, one can verify that

$$|w_1(n)| \leq |z_2|\frac{1+0.5\cos(\omega_0)}{1.25+\cos(\omega_0)} \approx |z_2|0.6772$$

for all $n \geq 0$. In particular, equality is achieved for $n = 0$.
z_1 is also periodic after the first 12 samples, given the periodicity of w_2 and w_1. Thus we have a family of candidates z_1 as function of z_2. We need now to select z_2 so that the constraint $\|(1/3)z_1\|_\infty \leq 1$ is satisfied. It is not difficult to verify that

$$z_2 = (1.25 + \cos(\omega_0))/(1 + 0.5\cos(\omega_0)) \approx -1.4766$$

guarantees the feasibility of z_2 and of the relative z_1 for Problem (6.29). Such solution is also the optimal dual solution, since the dual cost associated with this solution is

$$\frac{10}{3}z_{11} + z_2 = \frac{20}{3} - z_2(\frac{10}{3}0.6672 - 1) = 8.5234, \tag{6.31}$$

which is the same value achieved by the primal suboptimal solution. This immediately implies that 8.5234 is the optimal cost for Problem (6.28).

Optimal cost of Problem (6.25). To find the optimal value of Problem (6.25), once again, we provide feasible primal and (pre-)dual solutions and show that they have the same cost. If we consider the (pre-)dual of Problem (6.26), for any $z_1 \in c_0$ the resulting $V_2^* z_1$ is in c_0 too. This implies that, given any $\epsilon > 0$, there exists a N such that, if

$$\|V_2^* z_1 + z_2 \cos(t\omega_0)\|_\infty \leq 1,$$

then, for all $t \geq N$, $|z_2 \cos(t\omega_0)| \leq 1 + \epsilon$. This implies that, for any element $z_1 \in c_0$ (and therefore for any sequence with finite support), $|z_2 \cos(t\omega_0)|$ is constrained to have amplitude less or equal to 1 for all $t \geq 0$. Given the choice of ω_0, the maximum value of $|z_2|$ compatible with the constraints is 1.

It is not difficult to see that the optimal (pre-)dual cost is at least as big as the cost given by Equation (6.31) with $z_2 = -1$, which is equal to 7.9241. This is actually the optimal dual cost. To see this, consider the following feasible solution to Problem (6.26). Choose

$$q = 10\frac{1}{1 + 0.5\lambda}.$$

Then, $\Phi_1 = \frac{1}{3}q$, $\Phi_2 = 0$ and $\beta = -1.2574$ is a feasible solution with cost equal to

$$\left\|\frac{10}{3}\frac{1}{1+0.5\lambda}\right\|_1 + 1.2574 = 7.9241.$$

Therefore, we have that $\nu^o = 7.9241 < \mu^o = 8.5234$.

The above example shows that, in multi-block problems, the approximation scheme based on the truncation of the dual variables may not converge to the optimal cost for the original problem. It is also clear from the example presented that the gap is generated by the fact that an implicit constraint, on the amplitude of z_2, is generated when $V_2^* z_1$ is an element of c_0.

A promising approximation scheme is based on solving the following sequence of problems:

$$\begin{aligned}\mu_N = \quad & \inf_{\Phi \in \ell_1^{n_z \times n_w}} \quad \|\Phi\|_1, \\ \text{subject to:} \quad & \\ & \mathcal{A}_{feas}\Phi = b_{feas}, \\ & \mathcal{A}_{\mathcal{H}_\infty} T_N(\Phi) \leq \gamma 1\end{aligned}$$

where the rows of $\mathcal{A}_{\mathcal{H}_\infty} T_N$ are truncations, supported up to N, of the respective rows of $\mathcal{A}_{\mathcal{H}_\infty}$. The convergence properties of such scheme are under current investigation.

Summary and Comments. We have studied the optimal ℓ_1 control problem with added constraints on the magnitude of the closed loop frequency response where in particular, the frequency constraints are imposed at a finite number of frequency points. This problem has a primal-dual formulation with no duality gap, both in terms of convex optimization with infinite dimensional LMI constraints, and in terms of infinite dimensional LP. The main point we make in this chapter is however that having a primal-dual formulation with no duality gap does not automatically imply that we can compute converging primal and dual finite dimensional approximation. In particular, we have shown, using duality theory and an example, that the approximation method may fail to converge to the optimal cost of the original problem. This result has also direct negative implication for computation of lower bounds for the standard multi-block ℓ_1 problem with interpolations on the unit circle. In this case, the same derivation can show that FME method may fail to converge to the optimal ℓ_1 cost.

7. Mixed $\mathcal{H}_2/\ell_1$ Control

In this chapter, we study the problem of minimizing the $\mathcal{H}_2$ norm of the closed loop map between the exogenous inputs and the regulated outputs, keeping the ℓ_1 norm of the transfer function between possibly different inputs and outputs constrained below some level γ. Such design problem arises when good nominal rms-error performance resulting from white noise inputs is desired, and stability robustness ([36, 37]) against LTV causal perturbations of finite induced ℓ_∞-norm is to be maintained. Also, it arises when both white noise inputs and deterministic bounded amplitude inputs act on the system at the same time and a controller must be designed to minimize the output variance due to the white noise input and, at the same time, to keep the worst-case amplitude amplification of the deterministic input below some desired level.

The SISO case of this problem has been studied in [10], see also [38]. A computational method, based on Delay Augmentation has been presented in [39] for the general MIMO case, a method that does not require the computation of the zero interpolation condition can be found in [40].

We apply the results of Chapter 4 and show how both primal and dual problem can be approximated by finite dimensional convex optimization problems which provide converging upper and lower bounds to the optimal cost. Although we study in detail the $\mathcal{H}_2/\ell_1$ problem, similar results can be derived for the $\ell_1/\mathcal{H}_2$ problem using the same procedure.

7.1 Problem Statement

Before we formally describe the problem, we need to introduce some notation. Given the closed loop map, $\Phi \in \ell_1^{n_z \times n_w}$, the projection operator $\Pi_2 : \ell_1^{n_z \times n_w} \to \ell_1^{m_2 \times n_2}$, for some $m_2 \leq n_z$ and $n_2 \leq n_w$, allows us to select the sub-block of Φ for which the $\mathcal{H}_2$ norm of the impulse response must be minimized. The operator $\Pi_1 : \ell_1^{n_z \times n_w} \to \ell_1^{m_1 \times n_1}$ picks $m_1 n_1$ elements of Φ and rearranges them in a $m_1 \times n_1$ matrix of sequences. Π_1 is not defined as a projection to leave more flexibility in the selection of the part of the closed loop system whose ℓ_1 norm must be constrained.

Formally, the problem we want to solve is given by:

$$\begin{aligned} \mu^o = & \inf_{\Phi \in \ell_1^{n_z \times n_w}} \|\Pi_2 \Phi\|_2, \\ & \text{subject to:} \\ & A_{feas}\Phi = b_{feas} \\ & \|W_1 \Pi_1 \Phi W_2\|_1 \leq \gamma \end{aligned} \tag{7.1}$$

where W_1 and W_2 are diagonal matrices of dimension $m_1 \times m_1$ and $n_1 \times n_1$ respectively, and are used to scale the elements in $\Pi_1 \Phi$.

As posed, the problem is very general and might not correspond to a reasonable control problem. In particular, since Π_1 and Π_2 can pick any set of elements of the closed loop transfer function, the solution of the problem may not be in $\ell_1^{n_z \times n_w}$, i.e., the stability of the closed loop map may not be ensured. The following assumption prevents this to happen:

Assumption 7.1.1. *If $\Phi \in \ell_1^{n_z \times n_w}$ satisfies $A_{feas}\Phi = b_{feas}$ and $\|W_1 \Pi_1 \Phi W_2\|_1 \leq \gamma$, for some finite positive γ, then $\|\Phi\|_1 < \infty$.*

In other words, we assume that the elements of the closed loop map that have their ℓ_1 norm constrained are enough to force the whole closed loop map to be in $\ell_1^{n_z \times n_w}$. This assumption is reasonable and usually verified, and can always be imposed by selecting Π_1, W_1, and W_2 appropriately.

7.2 Dual Problem

The result of next theorem holds without Assumption (7.1.1) and unveils the structure of a dual problem with no duality gap. Once again, the next theorem and the one after it are applications of the results in Chapter 4. However, it is interesting to see how the selection of the various vector spaces and topologies is guided by the satisfaction of the hypothesis of the theorems in Chapter 4 and affects the information we extract from the primal-dual pair.

Theorem 7.2.1. *Given Problem (7.1), assume that γ is strictly greater than ν^o, the optimal cost of the associated standard ℓ_1 problem:*

$$\begin{aligned} \nu^0 = & \inf_{\Phi \in \ell_1^{n_z \times n_w}} \|W_1 \Pi_1 \Phi W_2\|_1 \\ & \textit{subject to:} \\ & A_{feas}\Phi = b_{feas}. \end{aligned}$$

Assume further that W_1^{-1} and W_2^{-1} exist. Then, the dual of Problem (7.1) with no duality gap is given by:

$$\begin{aligned} \mu^o = & \max \langle b_{feas}, z^* \rangle - \gamma\eta \\ & \textit{subject to:} \\ & A_{feas}^* z^* = \Pi_1^* w_1^* + \Pi_2^* w_2^* \\ & \|w_2^*\|_2 \leq 1, \ \|W_1^{-1} w_1^* W_2^{-1}\|_\infty \leq \eta \\ & \eta \in \mathbb{R}^+, \ z^* \in \ell_\infty, \ w_1^* \in \ell_\infty^{m_1 \times n_1}, \ w_2^* \in \ell_2^{m_2 \times n_2} \end{aligned} \tag{7.2}$$

Proof. To pose the problem in the standard form (4.1), we define the following cones in $X = \ell_1^{n_z \times n_w} \times \mathbb{R} \times \mathbb{R}$.

$$N_1 = \{\Phi, \xi_1, \xi_2, \mid \|W_1 \Pi_1 \Phi W_2\|_1 - \xi_1 \leq 0,\ \xi_1 \geq 0\}$$
$$N_2 = \{\Phi, \xi_1, \xi_2, \mid \|\Pi_2 \Phi\|_2 - \xi_2 \leq 0,\ \xi_2 \geq 0\}$$

Let $N = N_1 \cap N_2$ be the negative cone. Then,

$$\mu^o = \inf \ \xi_2$$
$$\text{subject to:}$$
$$\mathcal{A}_{feas}\Phi = b_{feas}$$
$$\xi_1 = \gamma$$
$$-[\Phi, \xi_1, \xi_2] \in P$$

where $P = -N$ is the positive cone. First, notice that the function $\|x\|_2$ is convex in ℓ_1. Furthermore, the cone N (and hence P) has a nonempty interior in X. The assumption that $\gamma > \nu^o$ ensures that there is a feasible solution in the interior of N.

Under the current assumptions, Theorem 4.2.3 can then be used to derive the dual problem and establish the lack of duality gap.

Although it may take a little work, it is left to the reader to verify that the conjugate cones of $P_1 = -N_1$ and $P_2 = -N_2$ are, respectively:

$$P_1^{\oplus} = \{\Phi^*, \xi_1^*, \xi_2^* \,|\Phi^* = \Pi_1^* w_1^*,\ \xi_1^* \leq 0,\ \xi_2^* = 0,$$
$$w_1^* \in \ell_\infty^{m_1 \times n_1},\ \|W_1^{-1} w_1^* W_2^{-1}\|_\infty + \xi_1^* \leq 0\}$$

and

$$P_2^{\oplus} = \{\Phi^*, \xi_1^*, \xi_2^* \mid \Phi^* = \Pi_2^* w_2^*,\ \xi_1^* = 0,\ \xi_2^* \leq 0,\ w_2^* \in \ell_2^{m_2 \times n_2},\ \|w_2^*\|_2 + \xi_2^* \leq 0\}.$$

We used the fact that Π_1 and Π_2 have closed range (being onto), in the derivation of the conjugate cones. The constraints given by the cone $P_1^{\oplus} + P_2^{\oplus}$ are now readily made explicit:

$$P_1^{\oplus} + P_2^{\oplus} = \{\Phi^*,\ \xi_1^*,\ \xi_2^* \mid \Phi^* = \Pi_1^* w_1^* + \Pi_2^* w_2^*,\ \xi_1^* \leq 0,\ \xi_2^* \leq 0,$$
$$w_1^* \in \ell_\infty^{m_1 \times n_1},\ w_2^* \in \ell_2^{m_2 \times n_2},$$
$$\|w_1^*\|_\infty + \xi_1^* \leq 0,\ \|w_2^*\|_2 + \xi_2^* \leq 0\}$$

It is left to the reader to verify that, after some rearrangements, the dual problem can be expressed as in Problem (7.2), which is the desired result. ■

The above theorem shows that an optimal solution of the infinite dimensional dual problem certainly exists in $\ell_\infty \times \ell_\infty^{m_1 \times n_1} \times \ell_2^{m_2 \times n_2}$. We do not know as yet if finite dimensional approximations of the dual problem are possible.

In order to avoid this unnecessary complication (that does not add much to the contribution of this chapter), we make the following assumption:

Assumption 7.2.1. *Any element, Φ_{ij}, of the closed loop map appears either in $\Pi_2 \Phi$ or in $\Pi_1 \Phi$, or in both.*

In practice, Assumption 7.2.1 can always be imposed by putting arbitrarily small weights, through W_1 and W_2, on the ℓ_1 norm of those Φ_{ij}'s that would not otherwise appear in either the $\mathcal{H}_2$ cost function or the ℓ_1 norm constraint. As shown in the next theorem, this assumption also guarantees that the optimal cost can be approximate arbitrarily well by feasible dual solutions, z, w_1, w_2, in $\ell_2 \times \ell_2^{m_1 \times n_1} \times \ell_2^{m_2 \times n_2}$, instead of $\ell_\infty \times \ell_\infty^{m_1 \times n_1} \times \ell_2^{m_2 \times n_2}$. The theorem also shows that, under Assumption 7.1.1, an optimal solution for problem (7.1) exists in $\ell_1^{n_z \times n_w}$.

Theorem 7.2.2. *Under Assumption 7.2.1, assume further that $\gamma < \nu^o$, and W_1^{-1} and W_2^{-1} exist. Let μ^o be the optimal cost of Problem (7.1). Then, the optimal cost is also given by the following optimization:*

$$\mu^o = \sup \quad \langle b_{feas}, z\rangle - \gamma\eta \tag{7.3}$$

subject to:

$$\mathcal{A}_{feas}^* z = \Pi_1^* w_1 + \Pi_2^* w_2$$
$$\|w_2\|_2 \leq 1, \; \|W_1^{-1} w_1 W_2^{-1}\|_\infty \leq \eta$$
$$\eta \in \mathbb{R}^+, \; z \in \ell_2 \; w_1 \in \ell_2^{m_1 \times n_1} \; w_2 \in \ell_2^{m_2 \times n_2}$$

and, under Assumption 7.1.1, a minimizing optimal solution to Problem (7.1) always exists.

Before we present the proof, we would like to add the following comment:

Comment: First, rewrite the dual problem in (7.2) as follows:

$$-\inf \quad \langle -b_{feas}, z\rangle + \gamma\eta \tag{7.4}$$

subject to:

$$\begin{bmatrix} \mathcal{A}_{feas}^* & -\Pi_1^* & -\Pi_2^* & 0 & 0 \\ 0 & 0 & 0 & 0 & 1 \end{bmatrix} \begin{bmatrix} z \\ w_1 \\ w_2 \\ \eta \\ \xi \end{bmatrix} = \begin{bmatrix} 0 \\ 1 \end{bmatrix}$$
$$\|w_2\|_2 \leq \xi, \; \|W_1^{-1} w_1 W_2^{-1}\|_\infty \leq \eta$$
$$\eta, \xi \in \mathbb{R}^+, \; z \in \ell_\infty, \; w_1 \in \ell_\infty^{m_1 \times n_1}, \; w_2 \in \ell_2^{m_2 \times n_2}$$

The existence of an optimal solution in $\ell_1^{n_z \times n_w}$ for Problem (7.1) would be proved almost directly if we could show that Problem (7.1) is itself a dual of Problem (7.2) rewritten as above with the variables $[z, w_1, w_2, \eta, \xi] \in {}^*X_1 = c_0 \times c_0^{m_1 \times n_1} \times c_0^{m_2 \times n_2} \times \mathbb{R}^2$. However, in this space, the cone

$$\{[z, w_1, w_2, \eta, \xi] \in {}^*X_1 \mid \|w_1\|_\infty - \eta \leq 0, \; \|w_2\|_2 - \xi \leq 0, \; \eta \geq 0, \; \xi \geq 0\}$$

has empty interior and hence the results in Theorem 4.2.3 do not apply.

Another possibility is to let $[z, w_1, w_2, \eta, \xi] \in {}^*X_2 = c_0 \times c_0^{m_1 \times n_1} \times \ell_2^{m_2 \times n_2}$. In this case, although the relative cone has a nonempty interior, the linear operator from ${}^*X_2 \to c_0^{n_z \times n_w} \times \mathbb{R}$,

$$\begin{bmatrix} A_{feas}^* & \Pi_1^* & \Pi_2^* & 0 & 0 \\ 0 & 0 & 0 & 0 & 1 \end{bmatrix} \tag{7.5}$$

does not have closed range. Thus, Theorem 4.2.3 cannot be applied to this case either.

Theorem 4.2.3 can be applied if we consider $[z, w_1, w_2, \eta, \xi] \in {}^*X = \ell_2 \times \ell_2^{m_1 \times n_1} \times \ell_2^{m_2 \times n_2}$, and consider the linear operator in (7.5) acting from ${}^*X$ to $\ell_2^{n_z \times n_w} \times \mathbb{R}$. Assumption 7.2.1 guarantees that this operator has closed range. In this case, the dual of this problem is Problem (7.1), but with $\Phi \in \ell_2^{n_z \times n_w}$. Thus, the optimal minimizing solution is ensured to exist in $\ell_2^{n_z \times n_w}$. However, it will be shown that, given Assumption 7.1.1, the optimal solution will be in $\ell_1^{n_z \times n_w}$. We are now ready to detail the proof of Theorem 7.2.2.

Proof. Consider Problem (7.3) rewritten as Problem (7.4) with $[z, w_1, w_2, \eta, \xi] \in {}^*X = \ell_2 \times \ell_2^{m_1 \times n_1} \times \ell_2^{m_2 \times n_2}$.

$$\underline{\mu} = \quad -\inf \quad \langle -b_{feas}, z\rangle + \gamma\eta$$

subject to:

$$\begin{bmatrix} \mathcal{A}_{feas}^* & -\Pi_1^* & -\Pi_2^* & 0 & 0 \\ 0 & 0 & 0 & 0 & 1 \end{bmatrix} \begin{bmatrix} z \\ w_1 \\ w_2 \\ \eta \\ \xi \end{bmatrix} = \begin{bmatrix} 0 \\ 1 \end{bmatrix}$$

$$\|w_2\|_2 \leq \xi,\ \|W_1^{-1} w_1 W_2^{-1}\|_\infty \leq \eta$$

$$\eta \geq 0, \xi \geq 0$$

$$[z, w_1, w_2, \eta, \xi] \in {}^*X$$

Let $\mathcal{A} : \ell_2^{n_z \times n_w} \times \mathbb{R} \to \ell_2 \times \ell_2^{m_1 \times n_1} \times \ell_2^{m_2 \times n_2} = X$ be

$$\mathcal{A} = \begin{bmatrix} \mathcal{A}_{feas} & 0 \\ \Pi_1 & 0 \\ \Pi_2 & 0 \\ 0 & 0 \\ 0 & 1 \end{bmatrix}.$$

Since X and $\ell_2^{n_z \times n_w} \times \mathbb{R}$ are reflexive and $\mathcal{A} : \ell_2^{n_z \times n_w} \times \mathbb{R} \to X$ is bounded, we have that ${}^*\mathcal{A} = \mathcal{A}^*$, i.e., the adjoint and the pre-adjoint of $\mathcal{A}$ are the same and equal to

$${}^*\mathcal{A} = \begin{bmatrix} \mathcal{A}_{feas}^* & \Pi_1^* & \Pi_2^* & 0 & 0 \\ 0 & 0 & 0 & 0 & 1 \end{bmatrix}.$$

We want to apply Theorem 4.2.3 to Problem (7.3). First, notice that $\mathcal{R}({}^*\mathcal{A})$ is closed in $\ell_2^{n_z \times n_w} \times \mathbb{R}$. To see this, it is enough to show that the operator

$$[\mathcal{A}_{feas}^* \quad \Pi_1^* \quad \Pi_2^*] : \ell_2 \times \ell_2^{m_1 \times n_1} \times \ell_2^{m_2 \times n_2} \to \ell_2^{n_z \times n_w}$$

has closed range. The closeness of $\mathcal{R}(\mathcal{A})$ will follow immediately.

From Assumption 7.2.1, and the definitions of Π_1 and Π_2, it follows that the operator $\begin{bmatrix} \Pi_1 \\ \Pi_2 \end{bmatrix}$ is 1-to-1, and it has closed range. Then, from Theorem

2.4.5 and Theorem 2.4.4, the adjoint, $[\,\Pi_1^* \quad \Pi_2^*\,]$, will be onto $\ell_2^{n_z \times n_w}$, and, therefore, the range of

$$[\,\mathcal{A}_{feas}^* \quad \Pi_1^* \quad \Pi_2^*\,] : \ell_2 \times \ell_2^{m_1 \times n_1} \times \ell_2^{m_2 \times n_2} \to \ell_2^{n_z \times n_w}$$

will be closed.

Now, define the following cones in ${}^*X$.

$$N_1 = \{z, w_1, w_2, \xi, \eta \in {}^*X \mid \|W_1^{-1} w_1 W_2^{-1}\|_\infty - \eta \le 0,\ \eta \ge 0\}$$
$$N_2 = \{z, w_1, w_2, \xi, \eta \in {}^*X \mid \|w_2\|_2 - \xi \le 0,\ \xi \ge 0\}$$

Let $N = N_1 \cap N_2$ be the positive cone. It is left to the reader to verify that N has a nonempty interior in ${}^*X$, and that the element $x = [0; 0; 0; 1; 1]^T$ satisfies ${}^*\mathcal{A}x = \begin{bmatrix} 0 \\ 1 \end{bmatrix}$ and belongs to the interior of N. The positive conjugate cones of $P_1 = -N_1$ and $P_2 = -N_2$ are given by:

$$P_1^{\oplus} = \{z^*, w_1^*, w_2^*, \xi^*, \eta^* \in X \mid z^* = 0, \|W_1 w_1^* W_2\|_1 + \eta^* \le 0, \eta^* \le 0, w_2^* = 0,\ \xi^* = 0\}$$
$$P_2^{\oplus} = \{z^*, w_1^*, w_2^*, \xi^*, \eta^* \in X \mid z^* = 0,\ w_1^* = 0,\ \|w_2^*\|_2 + \xi^* \le 0,\ \eta^* = 0,\ \xi^* \le 0\}$$

Finally, if $\gamma > \nu^o$, the cost of the dual Problem (7.2) is bounded. Since $\underline{\mu}$ is a lower bound for μ^o, then $\underline{\mu}$ is bounded. Therefore, Theorem 4.2.3 applies, and, after simple rearrangements, we have that

$$\underline{\mu} = \min_{\Phi \in \ell_2^{n_z \times n_w}} \|\Phi_2\|_2$$
$$\text{subject to:} \quad \mathcal{A}_{feas}\Phi = b_{feas}, \quad \|W_1 \Pi_1 \Phi W_2\|_1 \le \gamma.$$

But, from Assumption 7.1.1, it follows immediately that the optimal Φ must belong to $\ell_1^{n_z \times n_w}$. Thus $\mu^o = \underline{\mu}$. ∎

7.3 Finite Dimensional Dual Approximation

The result of the above theorem allows us to pose the dual problem in the separable space ${}^*X = \ell_2 \times \ell_2^{m_1 \times n_1} \times \ell_2^{m_2 \times n_2}$. The problem is still infinite dimensional in general. We want to derive, if possible, finite dimensional approximations of the dual Problem (7.3). In particular, we would like to replace the dual variable sequence z in ℓ_2 with a sequence of finite support, and study what happens as the support goes to infinity.

The fact that the dual problem is posed in a separable space is a good prerequisite if we want to derive finite dimensional approximations of the dual problem, but, unfortunately, it is not sufficient. A sufficient condition is the denseness of the approximating dual feasible solutions in the set of the dual feasible solutions. This guarantees that the cost of the approximate problem converges to the optimal dual cost, as the order of approximation increases.

Assumption 7.2.1 ensures that the situation described in the following example cannot happen.

Example. Consider the following set of stable closed loop maps

$$\Phi = \begin{pmatrix} \Phi_1 \\ \Phi_2 \end{pmatrix} = \begin{pmatrix} 1 \\ 0 \end{pmatrix} - \begin{pmatrix} 1-2\lambda \\ 1-5\lambda \end{pmatrix} Q, \quad Q \in \ell_1.$$

Since $(1-2\lambda)$ and $(1-5\lambda)$ are coprime, there are no zero interpolation conditions. The rank interpolation conditions are given in the λ domain by

$$(1-5\lambda)\hat{\Phi}_1 \quad - \quad (1-2\lambda)\hat{\Phi}_2 = 1-5\lambda.$$

Suppose we want to solve the following problem:

$$\mu^o = \inf_{\substack{\mathcal{A}_{feas}\Phi = b_{feas} \\ \|\Phi_1\|_1 \leq 2}} \|\Phi_1\|_2 \tag{7.6}$$

Note that, although the above problem does not satisfy Assumption 7.2.1, since Φ_2 is unconstrained, the result of Theorem 7.2.2 is still valid in this case due to the structure of the operator $\mathcal{A}_{feas}$.

The dual problem can be written as follows:

$$\mu^o = \sup \quad [1 \ \ -5 \ \ 0 \ \ \ldots] z - 2\eta$$

subject to:

$$\begin{bmatrix} 1 & -5 & 0 & 0 & 0 & \ldots \\ 0 & 1 & -5 & 0 & 0 & \ldots \\ 0 & 0 & 1 & -5 & 0 & \ldots \\ \vdots & \vdots & & \ddots & \ddots & \ddots \end{bmatrix} z + w_1 + w_2 = 0$$

$$\begin{bmatrix} -1 & 2 & 0 & 0 & 0 & \ldots \\ 0 & -1 & 2 & 0 & 0 & \ldots \\ 0 & 0 & -1 & 2 & 0 & \ldots \\ \vdots & \vdots & & \ddots & \ddots & \ddots \end{bmatrix} z = 0$$

$$\|w_2\|_2 \leq 1, \ \|w_1\|_\infty \leq \eta$$

$$\eta \in \mathbb{R}^+, \ z \in \ell_2 \ w_1 \in \ell_2 \ w_2 \in \ell_2$$

where $\mathcal{A}^*_{feas}$ and b_{feas} have been made explicit.

Notice that, if we force z to be of finite support, then the only feasible z is the zero sequence. Thus, the optimal cost, when $z \in FS$, is zero, while μ^o for this problem is equal to $\sqrt{3}/2$.

We must mention that the difficulty in this problem is due to the way we construct the operator $\mathcal{A}_{feas}$. In order to avoid this, we should have explicitly included the constraint $\hat{\Phi}_1(0.5) = 1$. In this simple case, the fix is easy. In more general examples, constructing the interpolation conditions to avoid such difficulties may be much more complicated, although we think it is always possible.

To simplify the statement of the next result, we introduce some notation. Given $\Psi \in \ell_2^{n_z \times n_w}$, with a slight abuse of notation, $\Pi_1 \Psi$ and $\Pi_2 \Psi$ denote the linear functionals acting on $\Pi_1 \Phi$ and $\Pi_2 \Phi$ respectively. Note that $\Pi_1 \Pi_1^* = I$, the identity operator from $\ell_2^{m_1 \times n_1}$ to $\ell_2^{m_1 \times n_1}$, and $\Pi_2 \Pi_2^* = I$, the identity operator from $\ell_2^{m_2 \times n_2}$ to $\ell_2^{m_2 \times n_2}$.

Let $\Psi = \Pi_1^* w_1 + \Pi_2^* w_2$. From Assumption 7.2.1, any element Ψ_{ij} of Ψ is either subject to the ℓ_∞ norm constraint, or to the ℓ_2 norm constraint, or both. Partition Ψ according to this classification. Let

$$
\begin{aligned}
\Psi^1 &= \{\Psi_{ij} \mid \Psi_{ij} \text{ is subject only to the } \ell_\infty \text{ norm constraint}\} \\
\Psi^2 &= \{\Psi_{ij} \mid \Psi_{ij} \text{ is subject only to the } \ell_2 \text{ norm constraint}\} \\
\Psi^{12} &= \{\Psi_{ij} \mid \Psi_{ij} \text{ is subject to both the } \ell_\infty \text{ and the } \ell_2 \text{ norm constraints}\}
\end{aligned}
$$

This partition induces analogous partitions on $\Pi_1 \mathcal{A}_{feas}^* z$, $\Pi_2 \mathcal{A}_{feas}^* z$, w_1, and w_2, denoted by

$$
\begin{aligned}
&\left[\Pi_1 \mathcal{A}_{feas}^* z\right]^1, \ [w_1]^1 \\
&\left[\Pi_2 \mathcal{A}_{feas}^* z\right]^2, \ [w_2]^2 \\
&\left[\Pi_1 \mathcal{A}_{feas}^* z\right]^{12}, [w_1]^{12}, [w_2]^{12}
\end{aligned}
$$

Note that $\left[\Pi_1 \mathcal{A}_{feas}^* z\right]^{12} = \left[\Pi_2 \mathcal{A}_{feas}^* z\right]^{12}$. Also, $\Pi_1 \Psi$ contains the union of Ψ^1 and Ψ^{12}. With this notation, the constraint $\mathcal{A}_{feas}^* z = \Pi_1^* w_1 + \Pi_2^* w_2$ in Problem (7.3) can be rewritten as follows:

$$
\begin{aligned}
\left[\Pi_1 \mathcal{A}_{feas}^* z\right]^1 &= [w_1]^1 \\
\left[\Pi_2 \mathcal{A}_{feas}^* z\right]^2 &= [w_2]^2 \\
\left[\Pi_1 \mathcal{A}_{feas}^* z\right]^{12} &= [w_1]^{12} + [w_2]^{12}.
\end{aligned}
\tag{7.7}
$$

Assume, without loss of generality, that the elements of Ψ in Ψ^{12} are in the top left corner of $\Pi_1 \Psi$, i.e., $(\Pi_1 \Psi)_{ij} \in \Psi^{12}$, for all $i = 1, \ldots, m$, and $j = 1, \ldots, n$, with $m \le m_1$ and $n \le n_1$. Then, using the first equation in (7.7), the ℓ_∞ norm constraint in Problem (7.3) can be rewritten as follows:

$$
\|W_1 w_1 W_2\|_\infty \le \eta \iff
$$

$$
\sum_{i=1}^{m} \sum_{k=m+1}^{m_1} \max \left\{ \begin{array}{c} \max_{1 \le j \le n} \rho_{ij} \| [w_1]_{ij}^{12} \|_\infty, \\ \max_{n+1 \le \ell \le n_1} \rho_{k\ell} \left\| \left[\Pi_1 \mathcal{A}_{feas}^* z\right]_{k\ell}^1 \right\|_\infty \end{array} \right\} \le \eta \tag{7.8}
$$

where ρ_{ij}'s and $\rho_{k\ell}$'s are the scaling factors due to the weighting matrices W_1 and W_2.

Finally, the ℓ_2 norm constraint in Problem (7.3) can be rewritten by combining the second and the third equation in (7.7), as follows:

$$\|w_2\|_2 \leq 1 \Leftrightarrow \left\| \left[\Pi_2 \mathcal{A}^*_{feas} z\right]^2 \right\|_2^2 + \left\| \left[\Pi_1 \mathcal{A}^*_{feas} z\right]^{12} - [w_1]^{12} \right\|_2^2 \leq 1 \qquad (7.9)$$

We can now state the next result.

Corollary 7.3.1. *Under the assumptions of Theorem 7.2.2 and Assumption 7.2.1, the optimal cost μ^o of Problem (7.3) is given by the following optimization:*

$$\mu^o = \sup \quad \langle b_{feas}, z\rangle - \gamma\eta \qquad (7.10)$$

subject to:

$$\sum_{i=1}^{m} \sum_{k=m+1}^{m_1} \max \left\{ \begin{array}{c} \max\limits_{1\leq j\leq n} \rho_{ij} \| [w_1]^{12}_{ij} \|_\infty, \\ \max\limits_{n+1\leq \ell\leq n_1} \rho_{k\ell} \left\| \left[\Pi_1 \mathcal{A}^*_{feas} z\right]^1_{k\ell} \right\|_\infty \end{array} \right\} \leq \eta$$

$$\left\| \left[\Pi_2 \mathcal{A}^*_{feas} z\right]^2 \right\|_2^2 + \left\| \left[\Pi_1 \mathcal{A}^*_{feas} z\right]^{12} - [w_1]^{12} \right\|_2^2 \leq 1$$

$$\eta \in \mathbb{R}^+, \ z \in FS \ [w_1]^{12} \in FS^{m\times n}$$

Moreover, a converging lower bound on μ^o can be computed by solving the above finite dimensional LMI optimization for increasing length of the support of z and $[w_1]^{12}$.

Proof. Substituting in Problem (7.3) for the constraints described in (7.8) and (7.9), we have that μ^o is given by Problem (7.10), but with $z \in \ell_2$ and $[w_1]^{12} \in \ell_2^{m\times n}$. Notice that, from Assumption 7.2.1, any element of $\mathcal{A}^*_{feas} z$ is constrained in norm. The result then follows from the boundedness of $\mathcal{A}^*_{feas}$ and the denseness of FS and $FS^{m\times n}$ in ℓ_2 and $\ell_2^{m\times n}$. Details are omitted.

For z and $[w_1]^{12}$ of finite support, the ℓ_2 norm constraint is a finite dimensional quadratic form in z and $[w_1]^{12}$. Finally, since the columns of the matrix representation of $\mathcal{A}^*_{feas}$ are elements of ℓ_1, for z of finite support $\mathcal{A}^*_{feas} z$ are decaying sequences. Using a standard argument for the ℓ_1 problem, it can be shown that only a finite number of the ℓ_∞ norm constraints are active in the optimization. It is actually possible to derive an upper bound on the number of active constraints. Thus, for z and $[w_1]$ of finite support, Problem (7.10) is equivalent to a finite dimensional convex optimization. ■

Note that a result similar to Theorem 5.3.2 holds for the approximation of the primal problem with finite support sequences. The details are omitted.

7.3.1 One-Block Problems

In this subsection, we specialize the results to one-block problems. One-block problems have special properties. For simplicity, we assume that $W_1 = I$, and $W_2 = I$.

Corollary 7.3.2. *Assume that Problem (7.1) is a one-block problem, and the constraint $\|\Pi_1\Phi\|_1 \le \gamma$ is active for each row of $\Pi_1\Phi \in \ell_1^{m_1\times n_1}$. Then, the elements of Φ constrained by $\|\Pi_1\Phi\|_1 \le \gamma$ have finite support.*

Proof. Without loss of generality, assume that $\Pi_1\Phi \in \ell_1^{m_1\times 1}$. In the one-block case, $x^* = A^*_{feas}z^*$ is an element of $c_0^{n_z\times n_w}$ for any $z^* \in \mathbb{R}^{c_z}$. With a slight abuse of notation, let $\Pi_1 x^*$ denote those elements of x^* in $c_0^{m_1\times 1}$ which act as linear functionals on the elements of Φ constrained by the ℓ_1 norm. From the structure of the dual constraints, it follows that $\Pi_1 x^* = w_1^* + \Pi_1\Pi_2^* w_2^*$, and, given the constraint on the norm of w_1^*, it also follows that $\|\Pi_1 x^* - \Pi_1\Pi_2^* w_2^*\|_\infty \le \eta$, with $\|w_2^*\|_2 \le 1$. The constraint $\|\Pi_1 x^* - \Pi_1\Pi_2^* w_2^*\|_\infty \le \eta$ can be written as follows: for each $i = 1,\ldots,m_1$, $\|(\Pi_1 x^* - \Pi_1\Pi_2^* w_2^*)_i\|_\infty \le \eta_i$, with $\eta_i \ge 0$ and $\sum_{i=1}^{m_1}\eta_i \le \eta$. Given that the ℓ_1 norm constraint is active for each row of $\Pi_1\Phi$, then, each η_i is strictly greater than *zero*. Since $x^* \in c_0^{n_z\times n_w}$ and $\Pi_2^* w_2^* \in \ell_2^{n_z\times n_w}$, only a finite number of the constraints imposed by $\|(\Pi_1 x^* - \Pi_1\Pi_2^* w_2^*)_i\|_\infty \le \eta_i$ are active in the optimization. From the alignment conditions we have that the elements of Φ constrained by the ℓ_1 norm have finite support. ■

In the special case of $\Pi_1 = \Pi_2 = I$, the dual of Problem (7.1) becomes:

$$\mu^o = \max \langle b_{feas}, z^*\rangle - \gamma\eta \quad (7.11)$$

subject to:

$$\|A^*_{feas}z^* + w_2^*\|_\infty \le \eta$$
$$\|w_2^*\|_2 \le 1$$
$$\eta \in \mathbb{R}^+,\ z^* \in \ell_\infty\ \ w_2^* \in \ell_2^{n_z\times n_w}$$

In this case, we have the following result that specializes the result of Corollary 7.3.2.

Corollary 7.3.3. *If Problem (7.1), with $\Pi_1 = \Pi_2 = I$, is a one-block problem, then, under the assumptions of Corollary 7.3.2, it has an optimal FIR solution.*

Proof. It follows immediately from Corollary 7.3.2 that the constraints of Problem (7.11) are inactive after the first N for some finite N. Problem (7.11) is therefore equivalent to a finite dimensional convex (LMI) problem. It is easy to see that the dual of such problem is finite dimensional too and is equivalent to Problem (7.1), with the variables $\Phi(k) = 0$ for $k \ge N$. The details are left to the reader. Thus, Problem (7.1) is intrinsically finite dimensional, and the optimal solution has finite support. ■

Remark 7.3.1. Notice that, if, in Theorem 7.2.1, we choose as primal space $\ell_2^{n_z\times n_w} \times \mathbb{R} \times \mathbb{R}$, instead of $\ell_1^{n_z\times n_w} \times \mathbb{R} \times \mathbb{R}$, the interior of the cone N will be empty. Thus, the existence of a separating hyperplane in $\ell_2^{n_z\times n_w} \times \mathbb{R} \times \mathbb{R}$ is not guaranteed by Theorem 4.2.2. We must point out that, if such hyperplane existed, then, it would be an element of $\ell_2^{n_z\times n_w} \times \mathbb{R} \times \mathbb{R}$. Given the structure of the dual problem (7.11), this would imply that the optimal solution is *FIR*.

This discussion raises doubts on the general existence of FIR optimal solutions to multiblock problems, even in the case $\Pi_1 = \Pi_2 = I$.

We conclude this chapter with an example.

7.3.2 Example

In the following example, we consider a filtering problem and show that the finite dimensional approximation of the dual problem is possible. Consider the

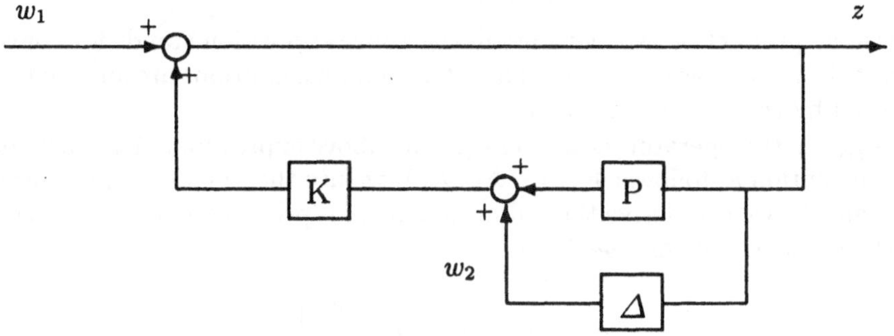

Fig. 7.1. Filtering Problem Block Diagram

closed loop system described by the block diagram shown in Figure 7.3.2. We want to design a controller K for the plant P (assumed stable), such that:

- the variance of the nominal error signal z ($\Delta = 0$) due to a white noise gaussian zero-mean input signal w_1 is minimized.
- the closed loop system is guaranteed to remain stable in presence of model uncertainty, represented as an additive perturbation Δ with $\|\Delta\|_{\ell_\infty - ind} <$ 1.

In other words, we want to find K such that $\|(1 - PK)^{-1}\|_2$ is minimized and $\|K(1 - PK)^{-1}\|_1 \leq 1$. The closed loop map from (w_1, w_2) to z is given by:

$$\Phi = \left((1 - PK)^{-1}, K(1 - PK)^{-1}\right).$$

The problem can be formally written as follows:

$$\begin{array}{rcc} \mu^o = & & \inf\limits_{K-stab} \quad \|\Phi_1\|_2 \\ & \text{subject to:} & \\ & & \|\Phi_2\|_1 \leq 1 \end{array}$$

Corollary 7.3.4. *If the ℓ_1 norm constraint is active, then the optimal Φ_2 is FIR.*

Proof. Since the plant is stable, we can use the Youla parametrization for stable plants, see [1], and obtain that $\Phi = (1 - PQ, -Q)$, with $Q \in \ell_1$, or, written in the usual form:

$$\Phi = H - QV = (1,0) - Q(P,1).$$

Since P and 1 are coprime, it follows from [41] (Lemma 4.1) that the set of the achievable closed loop maps is described by the set of all Φ satisfying the following constraints:

$$\Phi \begin{bmatrix} 1 \\ -P \end{bmatrix} = H \begin{bmatrix} 1 \\ -P \end{bmatrix},$$

where the product of two elements in the above expression stands for convolution of the respective sequences. The interpolation conditions are automatically satisfied by the above constraints.

$\mathcal{A}_{feas}$ is the operator acting on Φ in the above expression. $\mathcal{A}_{feas}$ acts on Φ by convolution as follows: $\Phi_1 * 1 - \Phi_2 * P$. If we use the Toeplitz representation for 1 and P, we can derive the following matrix representation of $\mathcal{A}_{feas}$ acting on the sequences Φ_1 and Φ_2 in ℓ_1:

$$\mathcal{A}_{feas}\Phi = [I \quad -\bar{P}] \begin{bmatrix} \Phi_1 \\ \Phi_2 \end{bmatrix},$$

where $\bar{P}$ is the Toeplitz matrix relative to P. Analogously, $b_{feas} = \mathcal{A}_{feas}H = [1, 0, 0, \ldots]^T$. Then, the feasibility constraints, $\mathcal{A}_{feas}\Phi = b_{feas}$, are given in matrix form by:

$$[I \quad -\bar{P}] \begin{bmatrix} \Phi_1 \\ \Phi_2 \end{bmatrix} = b_{feas}.$$

The problem we want to solve can now be posed as follows:

$$\begin{aligned} \mu^o = \quad & \inf \quad \|\Phi_1\|_2, \\ \text{subject to:} \quad & [I \quad -\bar{P}] \begin{bmatrix} \Phi_1 \\ \Phi_2 \end{bmatrix} = b_{feas} \\ & \|\Phi_2\|_1 \le 1 \end{aligned}$$

which is nothing but Problem (7.1). Its dual, Problem (7.2), assumes the following form:

$$\begin{aligned} \mu^o = \quad & \max \quad b^T z^* - \eta \\ \text{subject to:} \quad & \|z^*\|_2 \le 1 \\ & \|\bar{P}^* z^*\|_\infty \le \eta \\ & \eta \in \mathbb{R}^+,\ z^* \in \ell_\infty \end{aligned}$$

Note that the constraints now force the solution z^* to be in ℓ_2. Since the optimal $z^* \in \ell_2$, it follows that $\bar{P}^* z^* \in \ell_2$. Thus, only a finite number of constraints in $\|\bar{P}^* z^*\|_\infty \le \eta$ can be active for any fixed η. Since the problem satisfies Assumption 7.1.1, from Theorem 7.2.2, a minimizing solution for the primal problem exists in $\ell_1^{1\times 2}$. Finally, from the alignment conditions, it follows that the optimal Φ_2 is FIR. ■

Moreover, from Corollary 7.3.1, we can find finite support dual solutions to approximate arbitrarily closely the optimal cost μ^o by solving a finite dimensional convex optimization. The finite support dual solution of length N is computed by imposing extra constraints on the dual variable z_2^*, so that $z_2^*(n) = 0$ for $n > N$. Thus, as the length of the finite support approximation increases, the sequence of costs of the optimal finite support solutions converges from below to μ^o.

To apply these results to a concrete example, consider the first order plant

$$\hat{P}(\lambda) = \frac{0.2}{1 + 0.5\lambda}.$$

In this case, P is stable and has a stable inverse.

Notice that, from the structure of Φ, we can see immediately that $Q = 0$ would correspond to the optimal Q if we were minimizing the ℓ_1 norm of $K(1 - PK)^{-1}$ only. In this case, $\|K(1 - PK)^{-1}\|_1 = 0$. This implies that the closed loop will remain stable in presence of any perturbation Δ, with $\|\Delta\|_{\ell_\infty - ind} < \beta$, for any finite positive β. $Q = 0$ corresponds to the loop open situation, therefore we obtain no attenuation of the disturbance input. On the other hand, if we design the optimal rejector (in the $\mathcal{H}_2$ sense), of the white noise input w_1, we obtain that the optimal Q is given by $Q = P^{-1}$ (P has stable inverse). In this case, the optimal $\mathcal{H}_2$ norm of $(1 - PK)^{-1}$ equals 0, but $\|K(1 - PK)^{-1}\|_1 = \|-Q\|_1 = \|5(1 + 0.5\lambda)\|_1 = 7.5$. Thus, the stability robustness specification is not satisfied.

For the mixed $\mathcal{H}_2/\ell_1$ problem, we have that, for $N \geq 5$, the sequence of sub-optimal dual costs practically converges to the value 0.8083.

Since there is a feasible finite support solution for the primal problem, following an argument similar to the one given in Theorem 5.3.2, the FMV method can be used to compute FIR suboptimal primal solutions whose costs approximate from above the optimal cost μ^o. This is done by solving the convex optimization problem with extra constraints $\Phi_i(k) = 0$, for $k \geq N$ and $i = 1, 2$. The standard trick of writing $\Phi_2 = \Phi_2^+ - \Phi_2^-$, with $\Phi_2^+ \geq 0$ and $\Phi_2^- \geq 0$, allows to transform the convex but non-differentiable constraint due to the ℓ_1 norm into a set of linear constraints. Hence, the resulting optimization problem is the following:

$$\begin{aligned}\overline{\mu}_N^2 = \quad & \min \quad \Phi_1^T\Phi_1 \\ \text{subject to:} \quad & \begin{bmatrix} I & -\bar{P} & \bar{P}\end{bmatrix}\begin{bmatrix}\Phi_1\\ \Phi_2^+\\ \Phi_2^-\end{bmatrix} = b \\ & \begin{bmatrix} 0 & 1^T & 1^T\end{bmatrix}\begin{bmatrix}\Phi_1\\ \Phi_2^+\\ \Phi_2^-\end{bmatrix} \le \gamma \\ & \Phi_2^+ \ge 0,\ \Phi_2^- \ge 0 \\ & \Phi_2^+(k) = 0,\ \Phi_2^-(k) = 0, \text{ for } k \ge N\end{aligned}$$

where 1^T is an infinite row vector with all elements equal to 1.

It turns out that the optimal FIR solutions of length 10, 12 and 14 have costs $\bar{\mu}_{10}^o = 0.8086$, $\bar{\mu}_{12}^o = 0.8084$ and $\bar{\mu}_{14}^o = 0.8083$ respectively. Thus, μ^o is equal to 0.8083 within a tolerance of 10^{-4}.

Actually, a closer look at the sequence of suboptimal FIR solutions allows us to compute exactly the solution to the problem. For each $N = 8, 10, 12, 14$, the optimal Φ_2 is such that $\Phi_2(k) = 0$ for $1 \le k \le N-2$. $\Phi_2(0)$ and $\Phi_2(N-1)$ for each N are reported below:

N	$\Phi_2(0)$	$\Phi_2(N-1)$
8	−0.9922	−0.0078
10	−0.9981	−0.0019
12	−0.9995	−0.0005
14	−0.9999	−0.0001

Clearly, as N increases, the optimal Φ_2 converges to the sequence

$$\{-1 \quad 0 \quad 0 \quad \ldots\}.$$

Since $\Phi_2 = -Q$, the optimal Q for the mixed problem is equal to 1. The resulting optimal controller K is given by

$$K = -Q(1-PQ)^{-1} = -(1+P)^{-1}.$$

The optimal Φ_1 is equal to $1 - P$ with $\|\Phi_1\|_2 = 0.8083$.

Summary and Comments. The analysis of the solvability of the problems in the last three chapters has been rather complete. We have seen that there is often some freedom in selecting the setup of the optimization problem for a given multi-objective control problem. The choice of the space where the problem lies, the topology on the space, and the positive cone are important in determining the properties of the duality relationship of the associated primal-dual pairs, and in turn, affect our ability to derive exact or approximate solutions. We would like to point out that, although we have analyzed the convergence properties of the FMV-FME approximation methods, the approximation properties of other generic methods i.e., not tailored to a specific multiobjective problem, can also be analyzed using the general approach based on duality theory proposed in this monograph. We omit the analysis here since it will not add much to the contribution of this work. Some of these methods, which were derived for the ℓ_1 problem and that can be generalized to multi objective problems are described in the next chapter.

8. A New Computational Method for ℓ_1

8.1 Introduction

In the previous chapters we have extended the ℓ_1 control problem to multi-objective control problems. In this chapter, we utilize the solutions to mixed-objective problems to solve the ℓ_1 problem. We approximate the ℓ_1-objective with a mix of ℓ_1 and $\mathcal{H}_2$ norms. In this respect, the problem in not fundamentally different from the mixed $\ell_1/\mathcal{H}_2$ problem of the previous chapter. From the point of view of the application of duality theory, the key difference is that we dualize the problem written directly in terms of Youla parametrization (3.2), instead of dualizing the problem with the feasibility conditions as done in the previous chapters.

8.1.1 Computational Methods

The general multi-block ℓ_1 problem is equivalent to an infinite dimensional linear program. Several methods have been proposed in the literature to provide approximate solutions to this problem. The main ones are: Q-design, Finitely Many Variables - Finitely Many Equations approximation (FMV-FME), Delay Augmentation (DA), Convex programming approach using mixed $\ell_1/\mathcal{H}_\infty$, a geometric dynamic programming approach, and finally, a state space approach. Bellow is a brief account of the advantages and disadvantages of these methods.

The first approach, the Q-design method [7], is based on approximating the optimal controller by approximating the stable Q parameter (in the standard Youla parametrization) by an FIR system. The ℓ_1 problem then becomes a finite dimensional linear program in the parameters of Q. This provides an upper bound on the optimal solution which is guaranteed to converge to the actual optimal as the length of the FIR increases to infinity. The controller is automatically derived from the Q parameter. In this chapter, we will show that, for a large class of multi-block problems, this method guarantees that a subsequence of suboptimal solutions converges in norm to an optimal ℓ_1 solution. The disadvantage of this procedure is that it has no stopping criterion, and may cause order inflation in the controller due to the FIR approximation.

The FMV method is based on approximating the closed loop map by an FIR and thus has similar properties to the above method. The FME method is based on approximating the dual of the original ℓ_1 problem and that provides converging lower bounds of the optimal value. This resolves the issue of finding a stopping criterion, however, it requires computing interpolation conditions that characterize the closed loop map.

In the DA method, the multi-block problem is transformed into a one-block problem by introducing fictitious delayed inputs and outputs to the controller. This method has three important properties: 1) The convergence of the lower bound is generally fast, 2) there exists a particular ordering of the input and output channels that ensures the convergence of the upper bound together with the convergence in norm of the computed suboptimal solutions to an optimal ℓ_1 solution for all well posed multi-block problems, and 3) for many problems, there exists an ordering of the input and output channels that will generate a sequence of suboptimal controllers without order inflation. Although properties 2) and 3) are very interesting from a theoretical point of view, in practice, finding the right ordering can be difficult even for problems of moderate size in terms of number of input and output channels. The computation of a suboptimal controller, as well as the construction of the interpolation conditions, are the main practical limitations to the application of DA and, in general, of all the methods based on direct optimization on the space of closed loop maps. For details on the above methods, see [1, 16]. A method that avoids the computation of the interpolation conditions tightly related to the Q-design approach has been recently proposed in [42].

The geometric methods based on dynamic programming arguments [43] provide a recursive algorithm for computing FMV. It has a direct relation to the state space methods [18, 19, 44] as shown in [45] and in the next chapter. Nevertheless, their computational properties are still under investigations.

Finally, a mixed objective approach based on solving an $\ell_1/\mathcal{H}_\infty$ was suggested in [47]. The method provides converging suboptimal solutions, but does not provide a stopping criterion. The solutions are based on solving convex optimization problems.

The method presented in this chapter, see also [46], can be seen as a Q-design method based on a mixed objective optimization. It therefore provides directly computable suboptimal controllers, converging upper bounds to the optimal ℓ_1 cost, and norm convergence of the suboptimal solutions to the optimal ℓ_1 solution. Moreover, since the method is based on closed loop map approximation, it provides a converging lower bound to the optimal ℓ_1 cost.

The main idea can be better understood in the simple SISO case. Instead of minimizing the ℓ_1 norm of the closed loop impulse response, we minimize the square of the ℓ_1 norm of the first N samples plus the square of the $\mathcal{H}_2$ norm of the tail (after N) of the sequence. For $N = 0$ we have a standard $\mathcal{H}_2$ problem, and as $N \to \infty$ the problem becomes a standard ℓ_1 problem. This problem has a special property in that the solutions are achieved by FIR Q's of order $N-1$, when the Youla parametrization is derived from the optimal $\mathcal{H}_2$ controller. In order to solve the mixed objective optimization, we first solve the standard $\mathcal{H}_2$ problem. We then use the optimal $\mathcal{H}_2$ controller to compute

H,U, and V in the Youla parametrization ($\Phi = H - UQV$). Finally, we solve the convex optimization problem, which can be shown to be equivalent to a finite dimensional problem. We also discuss in detail all the convergence issues.

In particular, as the dimension of the quadratic programming problem increases, this method provides converging upper and lower bounds to the optimal ℓ_1 norm and, for well posed multi-block problems, ensures the convergence in norm of the suboptimal solutions to an optimal ℓ_1 solution. The new method does not require the computation of the interpolation conditions, and it allows the direct computation of the suboptimal controller.

8.2 Notation and Problem Setup

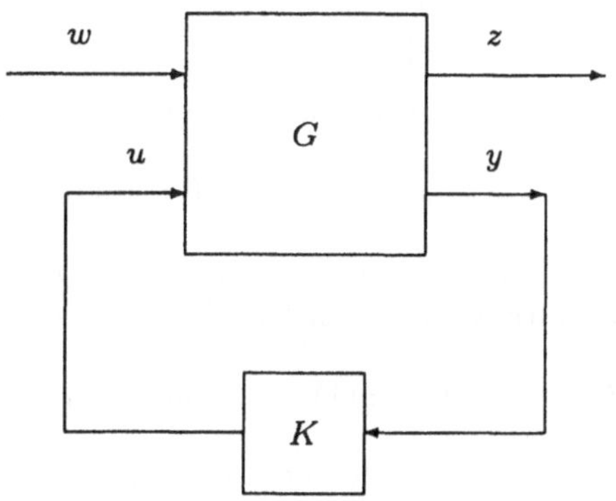

Fig. 8.1. General Set-Up

We consider the standard generalized system G shown in Figure 8.1

$$G = \left[\begin{array}{c|cc} A & B_1 & B_2 \\ \hline C_1 & D_{11} & D_{12} \\ C_2 & D_{21} & D_{22} \end{array}\right]$$

We make the following assumptions:

Assumption 8.2.1.

(A1) (A, B_2) *is stabilizable and* (C_2, A) *is detectable.*
(A2) D_{12} *has full column rank and* D_{21} *has full row rank.*
(A3) $\begin{bmatrix} A - Ie^{-i\theta} & B_2 \\ C_1 & D_{12} \end{bmatrix}$ *has full column rank for all* $\theta \in [0, 2\pi)$.
(A4) $\begin{bmatrix} A - Ie^{-i\theta} & B_1 \\ C_2 & D_{21} \end{bmatrix}$ *has full row rank for all* $\theta \in [0, 2\pi)$.
(A5) $D_{11} = 0$, *and* $D_{22} = 0$.

Assumptions $A1$ to $A4$ are standard assumptions, see [1, 48]. Assumption $A5$ is made only to simplify the derivations and can be removed.

We know from Section 3.2 that all stable closed loop maps from w to z can be parametrized as $\Phi = H - UQV$ where $Q \in \ell_1^{n_u \times n_y}$ is arbitrary, and H, U, and V are stable systems that can be computed from the problem data [1]. The ℓ_1 problem is given by:

$$\mu^o = \inf \|\Phi\|_1. \tag{8.1}$$

subject to:

$$\Phi = H - UQV$$
$$Q \in \ell_1^{n_u \times n_y}$$

Next, we are going to decompose the above expression in a special way. Let $\ell^{n\times m}$ be the space of all sequences of $n \times m$ matrices. For a fixed positive integer N, let $P_N : \ell^{n\times m} \to \mathbb{R}^{n\times m} \times \mathbb{R}^N$ denote the *truncation* operator of order N:

$$P_N x = [\, x(0) \quad x(2) \quad \dots \quad x(N-1)\,],$$

and $T_N : \ell^{n\times m} \to \ell^{n\times m}$ denote the *tail* operator of order N:

$$\cdot T_N x = [\, x(N) \quad x(N+1) \quad \dots]$$

. Also let $\bar{P}_N : \mathbb{R}^{n\times m} \times \mathbb{R}^N \to \ell^{n\times m}$ denote the operator that makes an infinite sequence from a finite sequence of length N by adding zeros.

$$\bar{P}_N y = [\, y(0) \quad y(1) \quad \dots \quad y(N-1) \quad 0 \quad 0 \quad \dots]$$

and $\bar{T}_N : \ell^{n\times m} \to \ell^{n\times m}$ denote the operator that puts N zeros at the beginning of a sequence

$$\bar{T}_N y = \Big[\underbrace{0 \quad 0 \quad \dots \quad 0}_{N} \quad y(0) \quad y(1) \quad \dots \Big]$$

Using these operators, Φ can be expressed as $\Phi = \bar{P}_N \Phi_1 + \bar{T}_N \Phi_2$ where $\Phi_1 = P_N \Phi$, $\Phi_2 = T_N \Phi$. Similarly $Q = \bar{P}_N P_N Q + \bar{T}_N T_N Q = \bar{P}_N Q_1 + \bar{T}_N Q_2$ and $H = \bar{P}_N P_N H + \bar{T}_N T_N H = \bar{P}_N H_1 + \bar{T}_N H_2$.

Also define:

$$U_1 \triangleq P_N U \bar{P}_N$$
$$U_{12} \triangleq T_N U \bar{P}_N$$
$$U_2 \triangleq T_N U \bar{T}_N$$

Note that $U_1 : \mathbb{R}^{n_u} \times \mathbb{R}^N \to \mathbb{R}^{n_z} \times \mathbb{R}^N$, $U_{12} : \mathbb{R}^{n_u} \times \mathbb{R}^N \to \ell_1^{n_z}$, and $U_2 : \ell_1^{n_u} \to \ell_1^{n_z}$. It follows from the Toeplitz structure of U that $U_2 = U$. Since U is causal, it follows that $U_{21} \triangleq P_N U \bar{T}_N = 0$.

Using these expressions, the decomposition of Φ can be written as:

$$\Phi_1 = H_1 - U_1 Q_1 V$$
$$\Phi_2 = H_2 - U_{12} Q_1 V - U Q_2 V$$

8.3 Approximation Method

Consider the following optimization problem:

$$\mu_N = \inf \sqrt{\|\Phi_1\|_1^2 + \|\Phi_2\|_2^2} \quad (8.2)$$

subject to:

$$\Phi_1 = H_1 - U_1 Q_1 V$$
$$\Phi_2 = H_2 - U_{12} Q_1 V - U Q_2 V$$
$$Q_1 \in \mathbb{R}^{n_u \times n_y \times N}, Q_2 \in \ell_1^{n_u \times n_y}$$

Properties of Problem (8.2)

(P1) For any N, denote

$$\|\Phi\|_N := \sqrt{\|\Phi_1\|_1^2 + \|\Phi_2\|_2^2}. \quad (8.3)$$

$\|\cdot\|_N$ is a norm on $X = \mathbb{R}^{n_z \times n_w \times N} \times \ell_2^{n_z \times n_w}$. Problem (8.2) is a constrained norm-minimization problem.

(P2) For each N the solution to Problem (8.2) exists.

(P3) There exists a parametrization $\Phi = H - UQV$, such that, for any finite N, Problem (8.2) is equivalent to the finite dimensional convex optimization

$$\mu_N = \inf \sqrt{\|\Phi_1\|_1^2 + \|\Phi_2\|_2^2} \quad (8.4)$$

subject to:

$$\Phi_1 = H_1 - U_1 Q_1 V$$
$$\Phi_2 = H_2 - U_{12} Q_1 V$$
$$Q_1 \in \mathbb{R}^{n_u \times n_y \times N}$$

Equivalently, the optimal solution with this parametrization satisfies $Q_2 = 0$.

Proof. (sketch) The first property is immediate. The second property follows from standard duality theory results, given the fact that Problem (8.2) is itself the dual of another convex optimization problem. For the third property, the special parametrization is the one obtained from the model-based optimal $\mathcal{H}_2$ controller. For that case, U and V are both inner. The details of the proof are in the Appendix. This property of the optimal $\mathcal{H}_2$ controller can also be derived from the principle of optimality in dynamic programming. ■

Next theorem gives the dual formulations for Problems (8.2) and (8.4). These formulations will be used in future sections.

Theorem 8.3.1. *The dual of Problem (8.2) with no duality gap is:*

$$\mu_N = \max \quad \langle H_1, x_1^* \rangle + \langle H_2, x_2^* \rangle \tag{8.5}$$

subject to:

$$U_1^* x_1^* V^* + U_{12}^* x_2^* V^* = 0$$
$$U_2^* x_2^* V^* = 0$$
$$\|x_1^*\|_\infty^2 + \|x_2^*\|_2^2 \leq 1$$
$$x_1^* \in \mathbb{R}^N,\ x_2^* \in \ell_2^{n_z \times n_w}$$

and the dual of Problem (8.4) with no duality gap is:

$$\mu_N = \max \quad \langle H_1, x_1^* \rangle + \langle H_2, x_2^* \rangle \tag{8.6}$$

subject to:

$$U_1^* x_1^* V^* + U_{12}^* x_2^* V^* = 0$$
$$\|x_1^*\|_\infty^2 + \|x_2^*\|_2^2 \leq 1$$
$$x_1^* \in \mathbb{R}^N,\ x_2^* \in \ell_2^{n_z \times n_w}$$

Proof. This is Theorem 1 in [22] pg.119. The details are left to the reader. ∎

8.4 Convergence Properties

In this section, we discuss the convergence of several quantities. First μ_N is shown to converge to μ^o. The solution to Problem 8.4 is a feasible solution to Problem 8.1 which has a cost denoted by $\bar{\mu}_N$. We will show that $\bar{\mu}_N$ converges to μ^o. In addition, we will derive a simple sequence of lower bounds on μ^o and show its convergence. Finally, we discuss the convergence of the actual solutions.

8.4.1 Convergence of the Cost μ_N

In the next theorem, we will show that the sequence μ_N converges to μ^o. The constraints in this problem can be rewritten as a function of Φ only by introducing a linear operator $\mathcal{A}_{feas}$, which is standard in the ℓ_1 literature. The problem is, then, equivalent to the following one:

$$\mu^o = \inf \quad \|\Phi\|_1 \tag{8.7}$$

subject to:

$$\mathcal{A}_{feas}\Phi = b_{feas}$$
$$\Phi \in \ell_1^{n_z \times n_w}$$

where $\mathcal{A}_{feas} : \ell_1^{n_z \times n_w} \to \ell_1$ is a linear bounded operator with closed range.

We can write two dual problems for the above minimization. One in $\ell_\infty^{n_z \times n_w}$, which is the norm-dual of $\ell_1^{n_z \times n_w}$, and the other in ℓ_∞, the dual of the constraint space. They are respectively given by:

$$\max \quad \langle H, x^* \rangle \tag{8.8}$$

subject to:

$$U^* x^* V^* = 0$$
$$\|x^*\|_\infty \leq 1$$
$$x^* \in \ell_\infty^{n_z \times n_w}$$

and

$$\max \quad \langle b_{feas}, z^* \rangle \tag{8.9}$$

subject to:

$$x^* = \mathcal{A}^*_{feas} z^*$$
$$\|x^*\|_\infty \le 1 \quad z^* \in \ell_\infty$$

It is worth mentioning that, all x^* satisfying $U^* x^* V^* = 0$ are in the range of $\mathcal{A}^*_{feas}$. We can now prove the following result.

Theorem 8.4.1. *Let μ_N be defined as in Problem (8.4). Then*

$$\lim_{N\to\infty} \mu_N = \mu^o$$

Proof. (sketch) From the optimal solution x^o, of the ℓ_1 problem, which exists under standard assumptions, we compute the sequence $\{\|x^o\|_N\}$, where $\|\cdot\|_N$ is defined as in (8.3). This provide a sequence convergent to μ^o with $\|x^o\|_N \ge \mu_N$ for any $N \ge 0$. Thus $\limsup_{N\to\infty} \mu_N \le \mu^o$.

To show that $\liminf_{N\to\infty} \mu_N \ge \mu^o$, we consider Problem (8.9), dual of Problem (8.7). It is well known that there are finite support feasible dual sequences z^* whose cost, μ, is arbitrarily close to μ^o from below. The main step is to show that from each z^* we can find feasible dual solutions to Problem (8.6) with cost approaching μ from below as N goes to infinity. The details are in the Appendix. ∎

8.4.2 Convergence of the Upper Bounds

For each N, the optimal solution of Problem (8.4) has $Q = [Q_1, 0] \in \ell_1^{n_u \times n_y}$ and thus the resulting optimal $\Phi^N = H - UQV = [\Phi_1^N \quad \Phi_2^N]$ is a BIBO stable closed loop map. Define

$$\overline{\mu}_N = \|\Phi^N\|_1$$

It follows that $\overline{\mu}_N \ge \mu^o$.

Next, we will show that $\overline{\mu}_N$ converges to μ^o as N goes to infinity. To do this, we first show that, if $\|\Phi_2^N\|_2$ is bounded, then $\|\Phi_2^N\|_1$ is also bounded. We then prove that the sequence $\|\Phi_2^N\|_2$ goes zero. To prove the first statement, we need to recall the following result:

Fact 8.4.1. *For the generalized system G there exists a parametrization of all closed loop stable maps $\Phi = H - UQV$ with polynomial $\hat{U}(\lambda)$ and $\hat{V}(\lambda)$.*

Lemma 8.4.1. *Consider the sequence of optimal solutions, $[\Phi_1^N \quad \Phi_2^N]$, of Problem (8.4), or equivalently of Problem (8.2), as a function of N. Then there exists a fixed M and two positive constants c_1 and c_2 such that*

$$\|\Phi_2^N\|_1 \le c_1 \|\Phi_2^N\|_2 + c_2, \qquad \text{for all } N \ge M.$$

Moreover, c_2 goes to zero as M approaches ∞.

Proof. For simplicity, We consider the case where $n_z = n$ and $n_u = n_w = 1$. Assume we have found a parametrization as in Fact 8.4.1. Consider Problem (8.2) with the resulting $\hat{U}(\lambda)$ polynomial. Note that, H may not be polynomial in general; however, it is an element of $\ell_1^{n\times 1}$ and hence of $\ell_2^{n\times 1}$.

Note that, Problem (8.2) is jointly convex in Q_1 and Q_2. This implies that, to find the optimal solution, we can first find the optimal Q_2 for each fixed Q_1 and then we can minimize with respect to Q_1. In our case, for a fixed Q_1, the optimal Q_2 is the one that minimizes the norm of Φ_2, i.e., the one that minimizes $\|H_2 - U_{12}Q_1 - UQ_2\|_2$. This is a standard $\mathcal{H}_2$ problem. However, instead of approaching it in the space $\mathcal{L}_2$ as done in the proof of P3 (see Appendix), it is convenient to look at it as a minimization on the space of one-sided sequences $\ell_2^{n_z\times n_w}(\mathbf{Z}_+)$. Thus, H_2 and Q_2 can be seen as sequences in $\ell_2^{n\times 1}(\mathbf{Z}_+)$, Q as an infinite block Toeplitz matrix, and Q_{12} as a matrix with infinite rows and N columns. Since this is a standard least squares minimization in an Hilbert space, the optimal Q_2 is given by:

$$Q_2 = U^+(H_2 - U_{12}Q_1)$$

and the optimal Φ_2^N is:

$$\Phi_2^N = (I - UU^+)(H_2 - U_{12}Q_1)$$

where U^+ denotes the pseudo-inverse of U. It is well known from the standard $\mathcal{H}_2$ optimization problem that the optimal Q_2 is in ℓ_1 and Φ_2^N is in $\ell_1^{n\times 1}$. If we denote by $A = I - UU^+$, then

$$\Phi_2^N = AH_2 - AU_{12}Q_1.$$

If $\hat{U}(\lambda)$ is a polynomial matrix, say of order $M-1$ then, for all the $N \geq M$, U_{12} has all zeros in the first $N - M$ columns. Moreover, the last M columns of U_{12} stay the same for all $N \geq M$ and their k elements are zeros for $k \geq M$. Thus, for $N \geq M$, the optimal Φ_2^N has a representation of the form:

$$\Phi_2^N = AH_2 + [0, B]Q_1$$

where $[0, B]Q_1 = AU_{12}Q_1$ and B maps $\mathbb{R}^M$ into $\ell_1^{n\times 1}$. In other words, Φ_2^N lies in a fixed translated subspace of $\ell_1^{n\times 1}$ and depends only on the last $M - 1$ elements of Q_1.

Consider now the sequence of optimal Q_1 for each N, denoted by Q_1^N. From the triangle inequality it follows that

$$\|\Phi_2^N\|_2 \geq -\|AH_2\|_2 + \|[0, B]Q_1^N\|_2,$$

or equivalently,

$$\|[0, B]Q_1^N\|_2 \leq \|\Phi_2^N\|_2 + \|AH_2\|_2.$$

From the fact that the range of B is fixed and finite dimensional in $\ell_1^{n\times 1}$ for each $N \geq M$, it follows that there exists a positive constant c_1 such that

$$c_1\|[0,B]Q_1^N\|_2 \geq \|[0,B]Q_1^N\|_1.$$

Therefore, we have that:

$$\|[0,B]Q_1^N\|_1 \leq c_1\|\Phi_2^N\|_2 + c_1\|AH_2\|_2$$

or,

$$\|\Phi_2^N\|_1 \leq c_1\|\Phi_2^N\|_2 + c_1\|AH_2\|_2 + \|AH_2\|_1$$

where we added and subtracted AH_2 in the norm on the left hand side, and we used the triangle inequality. Notice that, $\epsilon_N = c_1\|AH_2\|_2 + \|AH_2\|_1$ is a monotonically non increasing sequence; therefore, $\epsilon_N \leq \epsilon_M$ for all $N \geq M$. Besides, $\epsilon_M \to 0$ as $M \to \infty$. The result then follows if we let $c_2 = \epsilon_M$. ∎

The previous result together with next theorem determine the convergence of $\bar{\mu}_N$ to the optimal ℓ_1 cost.

Theorem 8.4.2. *Consider the sequence of optimal solutions to Problem (8.4) as a function of N, $[\Phi_1^N \quad \Phi_2^N]$. Then*

$$\lim_{N\to\infty} \|\Phi_2^N\|_2 = 0$$

Proof. See Appendix. ∎

As an immediate result form the above Theorem and Lemma 8.4.1 we have:

Corollary 8.4.1. *The sequence of upper bounds $\overline{\mu}_N$ converges to μ^o.*

Proof. For each N, let $\Phi^N = [\Phi_1^N, \Phi_2^N]$ be the optimal solution and denote μ_{1N} and μ_{2N} as follows:

$$\mu_{1N} = \|\Phi_1^N\|_1$$
$$\mu_{2N} = \|\Phi_2^N\|_2$$

Clearly, $\mu_N = \sqrt{\mu_{1N}^2 + \mu_{2N}^2}$. From Theorem 8.4.2 we have that $\mu_{2N} \to 0$ as $N \to \infty$. From Theorem 8.4.1 we have that $\mu_N \to \mu^o$ as $N \to \infty$. As consequence of these results, it follows that

$$\lim_{N\to\infty} \mu_{1N} = \mu^o.$$

From Lemma 8.4.1 it follows that $\|\Phi_2^N\|_1 \to 0$ as $N \to \infty$. Since it is always true that

$$\mu_{1N} \leq \overline{\mu}_N \leq \mu_{1N} + \|\Phi_2^N\|_1$$

We have that $\overline{\mu}_N \to \mu^o$ as $N \to \infty$. ∎

8.4.3 A Sequence of Lower Bounds

Next we derive an easily computable sequence of lower bounds for μ^o which will be denoted by $\underline{\mu}_N$. Assume, without loss of generality, that μ^o is strictly greater than 0; since $\mu^o = 0$ if and and only if $\mu_0 = \mu_{1N} = \mu_{2N} = 0$.

Consider the dual problem in (8.5). For a given N, $x^* = [x_1^*, x_2^*]$ are the dual variables. Let $\gamma_1^N = \|x_1^*\|_\infty$ and $\gamma_2^N = \|x_2^*\|_2$. From the alignment conditions it follows that

$$\begin{aligned}\gamma_1^N &= \frac{\mu_{1N}}{\mu_N}\\ \gamma_2^N &= \frac{\mu_{2N}}{\mu_N}.\end{aligned} \tag{8.10}$$

Given that $\|x_2^*\|_\infty \leq \sqrt{n_z}\|x_2^*\|_2$, it follows that the element $y^* = \dfrac{x^*}{\gamma_1^N + \sqrt{n_z}\gamma_2^N}$ has $\|y^*\|_\infty \leq 1$ and therefore, it is feasible for Problem (8.8), the dual of the ℓ_1 problem in (8.1). Hence,

$$\mu^o \geq \frac{\mu_N}{\gamma_1^N + \sqrt{n_z}\gamma_2^N}$$

Substituting the expressions in Equation (8.10) we obtain

$$\mu^o \geq \frac{\mu_N^2}{\mu_{1N} + \sqrt{n_z}\mu_{2N}} := \underline{\mu}_N.$$

Clearly $\underline{\mu}_N$ converges to μ^o as $N \to \infty$ since the numerator and the denominator converge to $(\mu^o)^2$ and μ^o respectively.

Remark 8.4.1. Notice that the results on the convergence of the upper and the lower bounds do not indicate that the sequences $\{\overline{\mu}_N\}$ and $\{\underline{\mu}_N\}$ are respectively monotonically non increasing and monotonically nondecreasing. In fact, they may not be as such, as shown in the example of Section 8.5.

8.4.4 Strong Convergence of the Suboptimal Solutions

We now address the convergence of the solutions of Problem (8.2), Φ^N. The results in this section are similar to the ones presented in [16].

Theorem 8.4.3. *For each N, let Φ^N be an optimal solution to Problem (8.2). Then, the sequence $\{\Phi^N\}$ contains a weak* convergent subsequence whose weak* limit is an optimal solution, Φ^o, for Problem(8.1). Moreover if the optimal solution is unique then the whole sequence converges to it.*

Proof. Since $\|\Phi^N\| = \overline{\mu}_N$ is a convergent sequence, $\{\Phi^N\}$ contains a *weak** converging subsequence. Denote the *weak** limit by Φ^{w^*}. For each N, Φ^N is feasible to Problem (8.1). This implies that Φ^N belongs to the set

$$\mathcal{S} = \{\Phi \,|\, \mathcal{A}_{feas}\Phi = b_{feas}\}.$$

However, S is *weak** closed, and thus it contains all its *weak** limit points. Therefore Φ^{w^*} is a feasible solution to Problem (8.1).

Φ^{w^*} is also an optimal solution to Problem (8.1). This follows from Lemma 2.5.5 in [16]:

$$\|\Phi^{w^*}\|_1 \le \liminf_{N_* \to \infty} \|\Phi^{N_*}\|_1 \le \mu^o.$$

If Φ^o is unique, then all subsequences must converge *weak** to it. Thus the whole sequence converges *weak** to Φ^o. ■

From the previous theorem, and theorem 2.4.3, it follows that, if any row of Φ^o, $(\Phi^o)_i$, achieves the optimal norm, μ^o, then the respective row of Φ^{N_*}, Φ^{N_*}, will converge in norm to $(\Phi^o)_i$.

Let $I \subset \{1, \ldots, n_z\}$ be the set of row indices for which $\|(\Phi^o)_i\|_1 = \mu^o$. $I = \{i \mid \|(\Phi^o)_i\|_1 = \mu^o\}$. Let $card(I)$ denote the cardinality of I, i.e. the number of elements in I. Given any $\Phi \in \ell_1^{n_z \times n_w}$ and any index set $I \subset \{1, \ldots, n_z\}$ with cardinality $card(I)$, we can construct $\Phi_I \in \ell_1^{card(I) \times n_w}$ by collecting only the rows on Φ whose index is in I. Define H_I and U_I analogously so that $\Phi_I = H_I - U_I Q V$.

Definition 8.4.1. *A multi-block problem with an optimal solution such that $card(I) = n_u$ is referred to as a well-posed problem.*

As noted in [16], for most multi-block problems the optimal solution achieves the optimal norm on at least n_u rows. However, it is possible to construct multiblock problems for which $n_I < n_u$. Such problems are not well-posed in the sense of Definition 8.4.1. We then have the following result:

Theorem 8.4.4. *Assume that Problem (8.1) is a well-posed multi-block problem. Let Φ^{N_*} be a subsequence of solutions to Problem (8.2) that converges weak* to Φ^o, an optimal solution to Problem (8.1) with $card(I) = n_u$. Assume further that $\hat{U}_I(\lambda)$ has full normal rank. Then $\|\Phi^{N_*} - \Phi^o\|_1 \to 0$ as $N \to \infty$.*

Proof. We have that $\Phi_I^{N_*}$ converges strongly to Φ_I^o. From the rank assumption on $\hat{U}_I$ we have that the map from $Q^{N_*}V$ to $\Phi_I^{N_*}$

$$\hat{Q}^{N_*}\hat{V} = \hat{U}_I^{-1}(\hat{H}_I - \Phi_I^{N_*})$$

is continuous with continuous inverse. Therefore, $Q^{N_*}V$ converges strongly to Q^oV. The result follows from the continuity in QV of the map $\Phi = H - UQV$. ■

Corollary 8.4.2. *The sequence Q^{N_*} converges strongly to Q^o.*

Proof. Since, by assumption $\hat{V}$ has full normal rank, there is a set J of n_y columns of $\hat{V}$ such that $V_J \in \ell_1^{n_y \times n_y}$ and $\hat{V}_J$ is invertible with continuous inverse. Given that $Q^{N_*}V$ is converging strongly to Q^oV then $Q^{N_*}V_J$ is also converging strongly to Q^oV_J. The result follows form the continuity of V_J^{-1}. ■

8.5 Example

In this section we briefly discuss some computational issues and present an example.

First, we describe how Problem 8.4 can be rewritten as a Linear Matrix Inequality problem. To avoid notational complications, we only describe the case where Φ is SISO. The generalization to the case where Φ is MIMO, although tedious, is straightforward, and is left to the reader.

We start by removing the variable Φ_1 and Φ_2 from the problem by rewriting it as follows:

$$\mu_N^2 = \inf \quad \gamma_1^2 + \|H_2 - U_{12}Q_1V\|_2^2. \tag{8.11}$$

subject to:

$$\|H_1 - U_1Q_1V\|_1 \leq \gamma_1$$
$$Q_1 \in R^N$$

Using the linearity in Q_1, U_1Q_1V can be rewritten in the following matrix form: $U_1Q_1V = A_f q_1$ where, $q_1 = [q_1(0), \ldots, q_1(N-1)]^T$ is the vector containing the first elements of the impulse response of Q_1, and $A_f \in R^{N\times N}$.

The ℓ_1 norm constraint can now be transformed into a set of linear constraints by a standard trick in linear programming. Namely, we have that

$$\left\{ \begin{array}{c} -\rho \leq H_1 - A_f q_1 \leq \rho \\ \rho(k) \geq 0 \\ A_{\ell_1} q_1 \stackrel{\Delta}{=} \sum_{k=0}^{N-1} \rho(k) \leq \gamma_1 \end{array} \right\} \Leftarrow \|H_1 - A_f q_1\|_1 \leq \gamma_1,$$

On the other hand, if $\|H_1 - A_f q_1\|_1 = \gamma_1$ then there exists a vector ρ such that

$$-\rho \leq H_1 - A_f q_1 \leq \rho$$
$$\rho(k) \geq 0$$
$$\sum_{k=0}^{N-1} \rho(k) = \gamma_1$$

Thus μ_n^2 is given by

$$\mu_N^2 = \inf \quad \gamma_1^2 + \|H_2 - U_{12}Q_1V\|_2^2. \tag{8.12}$$

subject to:

$$A_{\ell_1}\rho \leq \gamma_1$$
$$H_1 - A_f q_1 \leq \rho$$
$$-H_1 + A_f q_1 \leq \rho$$
$$\rho \geq 0,\ q_1 \in R^N$$

Now, consider the term $\|H_2 - U_{12}Q_1V\|_2^2$. From simple state space manipulations, $H_2 - U_{12}Q_1V$ can be represented as follows

$$H_2 - U_{12}Q_1V = \left[\begin{array}{c|c} \tilde{A} & \tilde{B}\begin{bmatrix} 1 \\ q_1 \end{bmatrix} \\ \hline \tilde{C} & 0 \end{array} \right]$$

Then

$$\|H_2 - U_{12}Q_1V\|_2^2 = Trace\left(\begin{bmatrix}1 & q_1^T\end{bmatrix}\tilde{B}^T L_o \tilde{B}\begin{bmatrix}1\\ q_1\end{bmatrix}\right)$$

where L_o is the observability gramian, which is given by unique positive definite solution of

$$\tilde{A}^T L_o \tilde{A} - L_o + \tilde{C}^T\tilde{C} = 0.$$

The dimension of L_o is fixed for a given problem and does not depend on N, the order of the approximation. This is because $H_2 - U_{12}Q_1V$ can be realized with $\tilde{A}$ and $\tilde{C}$ fixed and independent of N. Moreover, the computation of L_o is independent of q_1. Therefore, L_o can be pre-computed.

Finally, from [8], we have that the constraint

$$Trace\left(\begin{bmatrix}1 & q_1^T\end{bmatrix}\tilde{B}^T L_o \tilde{B}\begin{bmatrix}1\\ q_1\end{bmatrix}\right) \leq \gamma_2^2$$

is equivalent to the following LMI in X $(X = X^T)$, q_1, and γ_2^2:

$$Trace(X) \leq \gamma_2^2, \qquad \begin{bmatrix} X & \begin{bmatrix}1 & q_1^T\end{bmatrix}\tilde{B}^T \\ \tilde{B}\begin{bmatrix}1\\ q_1\end{bmatrix} & L_o^{-1}\end{bmatrix} \geq 0$$

Thus, Problem 8.4 can be solved by solving the following LMI problem:

$$\mu_N^2 = \inf \gamma^2. \tag{8.13}$$

subject to:

$$\begin{bmatrix}\gamma^2 - \gamma_2^2 & \gamma_1\\ \gamma_1 & 1\end{bmatrix} \geq 0$$

$$A_{\ell_1}\rho \leq \gamma_1$$

$$Trace(X) \leq \gamma_2^2$$

$$H_1 - A_f q_1 \leq \rho$$

$$-H_1 + A_f q_1 \leq \rho$$

$$\begin{bmatrix} X & \begin{bmatrix}1 & q_1^T\end{bmatrix}\tilde{B}^T \\ \tilde{B}\begin{bmatrix}1\\ q_1\end{bmatrix} & L_o^{-1}\end{bmatrix} \geq 0$$

$$\rho \geq 0,\ X = X^T,\ q_1 \in R^N$$

Note that, with a slight abuse of notation, we use $\geq$ and $\leq$ to describe both matrix inequalities and component-wise inequalities. Note however, that since the linear inequalities are a special case of matrix inequalities, they can be interpreted as component-wise matrix inequalities.

We now apply the new method to solve the ℓ_1 problem for the system described in Figure 8.2, where the plant is given by

$$\hat{P}(\lambda) = \frac{\lambda(\lambda - 0.5)}{(\lambda - 0.1)(1 - 0.5\lambda)}$$

and the weights are

$$\rho = 0.1; \quad \hat{W}_1(\lambda) = \tfrac{0.4}{1-0.6\lambda}; \quad \hat{W}_2(\lambda) = \tfrac{1-0.75\lambda}{1-0.25\lambda}$$

The same problem has also been considered in [16]. The closed loop system is

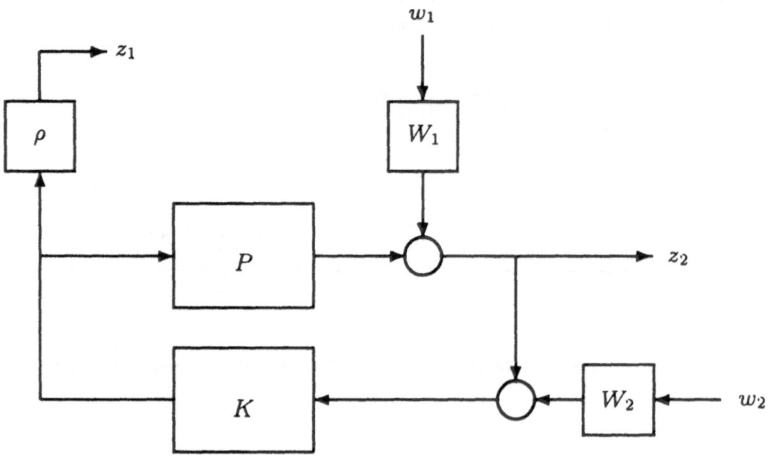

Fig. 8.2. System Configuration

the following

$$\Phi = \begin{pmatrix} \rho K(1-PK)^{-1}W_1 & \rho K(1-PK)^{-1}W_2 \\ (1-PK)^{-1}W_1 & PK(1-PK)^{-1}W_2 \end{pmatrix}$$

Figure 8.3 shows the convergence of the upper and the lower bounds for increasing order of $Q_1(\lambda)$. Notice that the sequence of upper bounds is not monotonically non-increasing.

For $N = 26$, which correspond to a polynomial Q_1 of order 25, we obtain that the suboptimal ℓ_1 solution Φ^N with $\overline{\mu}^N = \|\Phi^N\|_1 = 71.1147$. The maximum value of the lower bound is 71.0884. Thus $\|\Phi^N\|_1$ is within 0.04% of the optimal ℓ_1 norm, μ^o. The ℓ_1 norms of the single transfer functions are

$$\begin{bmatrix} \|\Phi_{11}\|_1 & \|\Phi_{12}\|_1 \\ \|\Phi_{21}\|_1 & \|\Phi_{22}\|_1 \end{bmatrix} = \begin{bmatrix} 1.8606 & 5.4428 \\ 26.0191 & 45.0956 \end{bmatrix}$$

From this information, we see that the second row is dominant in the problem and, in particular, Φ_{22} is the element with higher norm. This immediately

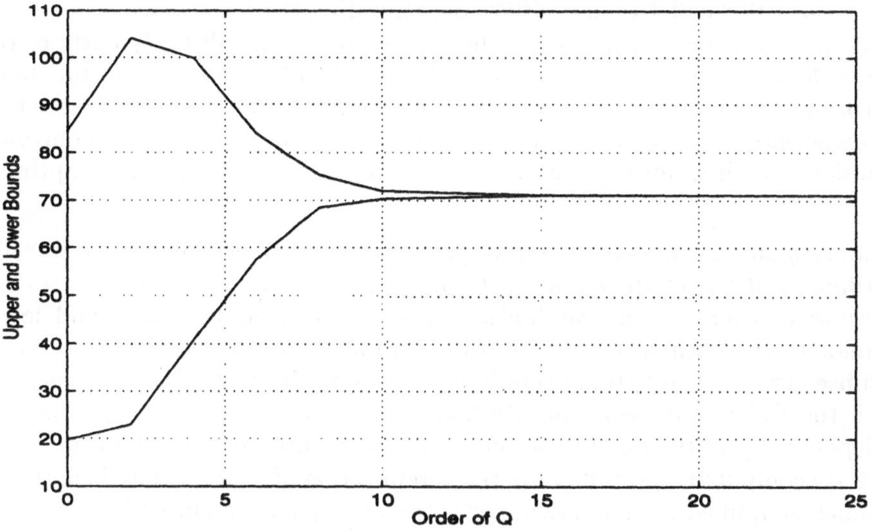

Fig. 8.3. Convergence of the Upper and Lower Bounds to the Optimal ℓ_1 Cost

indicates the right reordering for DA, namely, we need to switch the two outputs and the two inputs in order for the DA upper bound to converge to μ^o.

The coefficients of the suboptimal Q_1 for $N = 26$ are shown in Figure 8.4. Although $Q_1(k)$ is supported for all $N \leq 26$, the coefficients $Q_1(k)$'s for

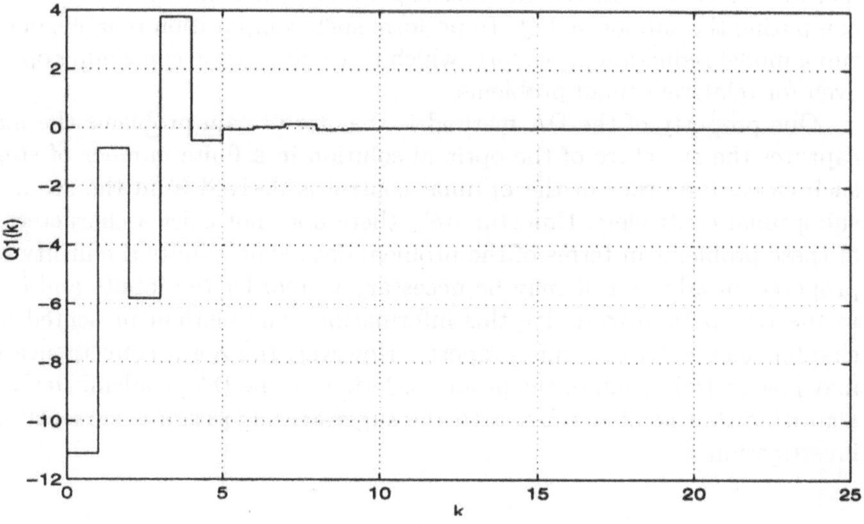

Fig. 8.4. Coefficients of Q_1 for $N = 26$

$k > 13$ are all smaller than $3 \cdot 10^{-5}$. By neglecting these coefficients we obtain that the suboptimal controller, K has order 16. The resulting ℓ_1 norm of the closed loop system, using such controller, is 71.1147. Thus, we do not loose much by considering Q_1 of order 12 instead of 25. If we only consider the first 9 coefficients of Q_1, as the figure suggests, we obtain a suboptimal controller of order 12 with ℓ_1 norm equal to 71.1296, still within the 0.06% of the optimal.

Summary and Comments. We have presented a new method to compute suboptimal solutions to the standard ℓ_1 problem. The new approach has several advantages over existing methods. It provides converging upper and lower bounds to the optimal ℓ_1 cost, and it guarantees the convergence in norm of a subsequence of suboptimal solutions to the optimal ℓ_1 solution . In contrast with the Delay Augmentation Method, the norm convergence is guaranteed independently of the inputs and outputs order. Moreover the suboptimal Q is directly computed, and therefore, the computation of the associated controller is much simplified in comparison with DA or any known closed loop approximation method. Another computational advantage is that the computations of the interpolation conditions are completely avoided.

For each N, we have to solve a semidefinite quadratic programming problem instead of a linear programming problem. The complexity of these two problems is not fundamentally different. While the number of constraints is approximately the same for the two problems, the number of variables in the new method is greater than the number of variables used by DA by approximately $n_u \times n_y \times N$ (the number of elements in Q). However, for a fair and realistic comparison of the complexity of the two methods, we must include the complexity of the computation of the interpolation conditions required by DA method. Moreover, for the DA method, we must also include the overhead of computing the suboptimal Q. To perform such computation reliably, one must run a model reduction procedure, which is usually a time consuming operation even for relatively small problems.

One property of the DA method is that for certain problems the method captures the structure of the optimal solution in a finite number of steps. In such cases, the order of the optimal control is derived from the sequence of suboptimal controllers. Unfortunately there does not exist a characterization of these problems in terms of the problem data which limits the utility of this property. In addition, it may be necessary to reorder the inputs and outputs in the DA method to derive this information. The method presented in this chapter does not poses this property. However, the norm convergence result may give an indication to the proper ordering for the DA problem. Deriving an algorithm that combines DA with the suggested algorithm is currently under investigation.

9. Nonlinear Controllers for Minimizing the Worst-Case Peak to Peak Gain

In this chapter, we depart from computational methods for LTI systems that provide LTI controllers, and consider approaches that can possibly lead to nonlinear controllers. The minimization of commonly used performance criteria, such as $\mathcal{H}_2$ or $\mathcal{H}_\infty$, automatically results in a LTI optimal controller when the plant is LTI. However, this is not necessarily the case for other performance criteria. In general, we are interested in nonlinear controllers either because they can achieve indeed better performance than the optimal LTI ones, or because they may be easier to implement or to compute.

As described next, both these motivations are valid in the case of minimization of the worst-case peak to peak gain. One characteristic of the optimal linear ℓ_1 controllers is that they can be of arbitrarily high order. This happens even in the full state feedback case. As shown in [49], the optimal linear full state feedback controller can be dynamic of very high order. This results have motivated the search for nonlinear controllers, hopefully less complex, that achieve the same performance as a given linear controller [18, 44]. Further motivation to look for the optimal nonlinear worst-case peak to peak controller has been given by the result in [17]. In that paper, it is shown that the optimal worst-case peak to peak gain achievable with nonlinear controllers is strictly better than the one achievable with linear controllers. In [19], two main contributions for the nonlinear full state feedback problem are presented. The first is that the optimal nonlinear full state feedback is static. The second is an algorithm that allows to check if a certain level of performance is achievable. The bisection algorithm, standard in the $\mathcal{H}_\infty$ literature, can then be used to obtain increasingly better suboptimal costs and suboptimal controllers. The results in [19], (see also [50] for some generalization) relies heavily on viability theory.

Here, we propose a more direct and natural approach to the full state feedback problem based on dynamic programming. The roots of this approach go back to the work in [51]. We first study the finite horizon problem with terminal state cost. We formulate two problems: the optimality problem and the feasibility problem. In the first one, we want to obtain the optimal cost and the optimal full state feedback strategy, when the input is arbitrary, unknown, but of bounded known amplitude. In the second one, we are only interested in checking if a given performance level is achievable, given that the input is arbitrary, unknown, but of bounded known amplitude.

As in [52], we look at the dual formulation of the dynamic program and, under mild assumptions, we show that it is possible to characterize the solutions for each finite horizon problem in terms of special polytopic gauge functions. The nice feature of the polytopes characterizing these functions is that they can be recursively computed using simple fixed transformations involving only the vertices of the polytopes.

We further show that, for a particular selection of the terminal cost, the polytopes generated at each step of the feasibility problem are exactly the polytopes generated by the invariant kernel algorithm used in [19]. However, the standard computational advantage of dynamic programming allows to discover and discard a whole set of redundant vertices that are instead considered in the algorithm proposed in [19]. Thus, the invariant kernel algorithm is strongly related to the cost to go of finite horizon feasibility problem with a special terminal cost. Furthermore, if the feasibility problem admits a solution, the invariant set to which it converges is the set of states for which the given performance is achievable on the infinite horizon. An immediate consequence of this equivalence is that the method providing a stopping criterion given in [19] also applies to the dynamic programming approach.

Using the same terminal cost function, we can alternatively solve the optimality problem. As already mentioned, this method provides the exact optimal cost for any finite horizon and it does not require a gamma iteration scheme. In certain cases, the polytopes stop changing after a finite number of iterations. In these cases, the optimality problem provides the exact infinite horizon optimal cost and optimal strategy. In general, however, the method provides converging lower bounds to the optimal infinite horizon cost.

Given that the approach presented in [19] is a special case of the dynamic programming approach proposed here, it is possible to use the method developed in [19] to derive internally stabilizing suboptimal controllers. We also present a different, computationally more efficient, method to derive upper bounds and suboptimal solutions. For a given step of the recursion, we can compute a suboptimal controller and a measure of the accuracy of the achieved cost with respect to the optimal one. It is important to have good upper bounds to reduce the computational complexity. A loose upper bound means that we may have to go through many iterations before achieving the desired guaranteed accuracy. Given the combinatorial computational complexity associated with each iteration step, it is desired to limit the number of iterations as much as possible.

The chapter is organized as follows: The notation and some mathematical preliminaries are in Section 9.1. Section 9.2 contains the problem setup. In Section 9.3 we present the main results for the finite horizon optimality and feasibility problems. We also describe the structure of the optimal strategy for the optimality problem. Section 9.4 contains infinite horizon generalization. It also presents a detailed comparison of this approach with the one presented in [19], and describes how to obtain suboptimal stabilizing controllers. Section 9.5 contains some examples. Although the treatment in this chapter is rather complete and self contained, the research in this field is at a preliminary stage. The results obtained hint to nontrivial connections with some neural network control architecture, and lead toward more general methods of approximate dynamic programming. Some suggestions for future work, are outlined in Section 9.5.2.

9.1 Notation and Preliminaries

In this section, we briefly define some notation and present preliminary results and definitions. A very good reference for the material presented in this section is [53].

In this chapter we consider convex, balanced, and compact polytopes in finite dimensional spaces. we recall that a set P is balanced if $x \in P$ implies that $-x \in P$. Given a convex balanced polytope P in a finite dimensional space, the Gauge function of P is given by

$$J(x) = \min_{x \in \gamma P} \gamma$$

Notice that the value of $J(x)$ can be infinite.

The polar of P, denoted by P^o, is defined as follows:

$$P^o = \{p \in R^n, \mid p^T x \leq 1,\ \forall x \in P \subset R^n\}.$$

The polar of P allows us to write a dual representation of $J(x)$. Namely,

$$J(x) = \max_{p \in P^o} p^T x$$

For a given polytope P, *vertP* denotes the set of its (external) vertices. P is completely defined by its vertices, since it is given by the convex hull of these vertices:

$$P = conv(vertP).$$

If $p_1, \ldots, p_l$ are all the vertices of P, we denote with V_P the matrix whose rows are the vertices of P, namely,

$$V_P = \begin{bmatrix} p_1^T \\ p_2^T \\ \vdots \\ p_l^T \end{bmatrix}$$

A special polytope is the unit ℓ_∞-ball. We denote it with

$$\mathcal{B} = \{x \in R^n, \mid \|x\|_\infty \leq 1\}.$$

Notice that $\mathcal{B}^o$ is the unit ℓ_1-ball.

In the rest of the chapter we will use the symbol $\mathcal{B}^o$ to denote the ℓ_1-ball in spaces of different dimensions. The dimension of the space will be context dependent. Sometimes, $\mathcal{B}^o$ will also be called the unit diamond. From the previous definitions, we have that

$$\|x\|_\infty = \min_{x \in \gamma\mathcal{B}} \gamma = \max_{p \in \mathcal{B}^o} p^T x$$

9.2 Problem Setup

We consider discrete time, linear, MIMO systems of the form

$$\begin{aligned} x_{k+1} &= Ax_k + B_1 w_k + B_2 u_k \\ z_k &= C_1 x_k + D_{11} w_k + D_{12} u_k \\ y_k &= C_2 x_k + D_{21} w_k \end{aligned}$$

The system has n_w exogenous inputs, n_u control inputs, n_y measurements outputs, and n_z regulated outputs. For each k, w_k is the vector of exogenous inputs, with the property that $\|w_k\|_\infty \leq \delta$ for some positive δ. The general problem is to select the control input u_k based only on the measurements y_k, $u_k = F_k(y_k)$ so that

- the closed loop system is asymptotically stable
- the maximum amplitude of z_k for all k is minimized.

In this chapter, we restrict our attention to the full state feedback problem ($C_2 = I$ and $D_{21} = 0$).

To simplify the exposition, we assume that $D_{11} = 0$. The results obtained for this case will extend naturally to the case where $D_{11} \neq 0$. We also assume that C_1 has rank n_s. This last assumption is needed only for the infinite horizon problem. It allows us to recover Shamma's results [19]. This assumption simplifies some convergence proofs for the infinite horizon. Research is ongoing to relax this assumption.

We first consider the finite horizon problem.

9.3 Finite Horizon Full State Feedback

In the finite horizon problem, we drop the requirement of asymptotic stability, and we consider a slightly more general cost function by introducing a terminal cost.

Consider the finite horizon $T = [0, \ldots, N-1]$. Denote with $u_{[0:N-1]} \in R^{n_u \times N}$, $w_{[0:N-1]} \in R^{n_w \times N}$ and $z_{[0:N-1]} \in R^{n_z \times N}$, the respective sequences over the horizon. Assume zero initial conditions, $x_0 = 0$, and consider the following optimization problem with weight on the final state, x_N:

$$\gamma^o = \inf_{u_{[0:N-1]}\in R^{n_u\times N}} \max_{\|w_{[0:N-1]}\|_\infty \le \delta} \max\left\{\|z_{[0:N-1]}\|_\infty, J_0(x_N)\right\} \tag{9.1}$$

where $J_0(x)$ is the Gauge function of a compact polytope Q_0. In the dual representation, we have that

$$J_0(X) = \sup_{q\in Q_0^o} q^T x.$$

Problem (9.1) is equivalent to a linear program. However, we are not as much interested in the optimal sequence $u_{[0:N-1]}$ as we are in the optimal strategy $F_k(x_k)$. This motivates the dynamic programming approach we consider in this chapter.

Using the principle of optimality, the above static minmax problem can be transformed into a differential minmax problem as follows. Define

$$J_1(x,\delta) = \min_{u_{N-1}} \max_{\|w_{N-1}\|_\infty \le \delta} \max\left\{\|z_{N-1}\|_\infty, J_0(Ax + B_1 w_{N-1} + B_2 u_{N-1})\right\},$$

where u_{N-1}, w_{N-1} and z_{N-1} are the last elements of the sequences $u_{[0:N-1]}$, $w_{[0:N-1]}$, and $z_{[0:N-1]}$ respectively. $J_1(x,\delta)$ gives the minimal cost we pay if we are at state x one step before the end of the game. Knowing J_1, we can similarly define

$$J_2(x,\delta) = \min_{u_{N-2}} \max_{\|w_{N-2}\|_\infty \le \delta} \max\left\{\|z_{N-2}\|_\infty, J_1(Ax + B_1 w_{N-2} + B_2 u_{N-2}, \delta)\right\}$$

$J_2(x,\delta)$ is the minimal cost we pay if we are at state x two steps before the end of the game. Thus, we have the following Bellman equation:

$$J_{k+1}(x,\delta) =$$
$$\min_{u_{N-(k+1)}} \max_{\|w_{N-(k+1)}\|_\infty \le \delta} \max\left\{\begin{array}{c}\|z_{N-(k+1)}\|_\infty, \\ J_k(Ax + B_1 w_{N-(k+1)} + B_2 u_{N-(k+1)}, \delta)\end{array}\right\}$$

Notice that, in this way, the value of Problem (9.1) is given by $J_N(0,\delta)$.

9.3.1 The optimality Problem (The Scalar Disturbance Case)

In this subsection, to simplify the exposition, we further assume that the system has only one exogenous input channel; B_1 is a column vector. The result obtained for this case is easily generalizable. The following definition simplifies the statement of the next Theorem.

Definition 9.3.1. *Given two matrices $F \in R^{l\times n}$ and $G \in R^{l\times n+1}$, $G = [G_1, G_2]$ with $G_1 \in R^{l\times n}$ and $G_2 \in R^{l\times 1}$, define the linear operator $\mathcal{T} : R^{l\times n} \times R^{l\times n+1} \to R^{3l\times 3n+1}$ as follows:*

$$\mathcal{T}(F,G) = \begin{bmatrix} F & 0 & 0 & 0 \\ 0 & G_1 & 0 & G_2 \\ 0 & 0 & G_1 & G_2 \end{bmatrix}$$

The following theorem shows that $J_N(x, \delta)$ is a Gauge function of some polytope Q_N. It also provides a recursive expression of the polar sets Q_k^o in terms of Q_{k-1}^o, for all $k = 2, \ldots, N$.

Theorem 9.3.1. *Let Q_0 be a compact polytope in R^{n_s}, with n_s being the dimension of the system state space. Let R_0 be the unit ℓ_∞-ball in R^{n_z}.*

For $k = 1$, define

$$Q_1^o = \tilde{A}^T \left\{ p \in S_1^o \,|\, p^T \tilde{B} = 0 \right\},$$

where

$$\tilde{A} = \begin{bmatrix} C_1 & 0 \\ A & B_1 \\ A & -B_1 \end{bmatrix}, \qquad \tilde{B} = \begin{bmatrix} D_{12} \\ B_2 \\ B_2 \end{bmatrix}$$

and

$$S_1^o = \begin{bmatrix} V_{R_0^o}^T & 0 & 0 \\ 0 & V_{Q_0^o}^T & 0 \\ 0 & 0 & V_{Q_0^o}^T \end{bmatrix} \mathcal{B}^o.$$

For $2 \le k \le N$, define

$$Q_k^o = \overline{A}^T \left\{ s \in S_k^o \,|\, s^T \overline{B} = 0 \right\},$$

where

$$\overline{A} = \begin{bmatrix} C_1 & 0 \\ A & B_1 \\ A & -B_1 \\ 0 & I \end{bmatrix}, \qquad \overline{B} = \begin{bmatrix} D_{12} \\ B_2 \\ B_2 \\ 0 \end{bmatrix}$$

and S_k^o is given by

$$S_k^o = \mathcal{T}(V_{R_0^o}, V_{Q_{k-1}^o})^T \mathcal{B}^o.$$

Then, for any $k \in [1, \ldots, N]$,

$$J_k(\chi, \delta) = \max_{q \in Q_k^o} q^T \begin{bmatrix} \chi \\ \delta \end{bmatrix}.$$

Proof. See Appendix. ∎

We must point out that the result of this theorem allows us to solve Problem (9.1) exactly ($\gamma^o = J_N(0, \delta)$). The computation of the sets Q_k follows simple transformation rules.

Notice that the problem we are solving can be reinterpreted as a full information problem. The same approach can be applied in cases where the model for the disturbance is different from the worst-case bounded amplitude one. In particular, a similar procedure could be derived for fixed input disturbances such as steps, sinusoids etc. Also specific constraints on the amplitude of the control input, such as actuator saturation constraints, can be easily added to the formulation. Clearly, in this cases, we expect more complex transformations in the set recursion.

9.3.2 Structure of the Optimal Controller

Theorem 9.3.1 also allows us to uncover the structure of the optimal controller. In fact, from the intermediate steps of the proof of the theorem, we have that, for $N > 1$:

$$J_N(\chi,\delta) = \min_u \max_{s \in S''_N} s^T \left[\overline{A} \begin{bmatrix} \chi \\ \delta \end{bmatrix} + \overline{B}u \right]$$

This expression is equivalent to the following minimization problem, which is equivalent to a linear programming problem:

$$J_N(\chi,\delta) = \min_u \left\| V_{S''_N} \overline{A} \begin{bmatrix} \chi \\ \delta \end{bmatrix} + V_{S''_N} \overline{B}u \right\|_\infty \tag{9.2}$$

An analogous expression holds for $N = 1$. Thus, for each $k \in [0, \ldots, N]$ and a given δ, the optimal control input u_k is obtained by solving a linear programming problem function of the state x_k.

Note that the optimal strategy for the finite horizon problem is, in general, time varying. The linear program that must be solved is in general different at each step.

Note also that the minimizing u may not be unique. To see how this may happen, assume that some of the elements of $V_{S''_N} \overline{B}$ are equal to zero. Without loss of generality, let such elements be the first l. $V_{S''_N} \overline{B}$ can be partitioned as follows:

$$V_{S''_N} \overline{B} = \begin{bmatrix} 0 \\ B_{nz} \end{bmatrix}.$$

Partition $V_{S''_N} \overline{A}$ analogously, i.e.,

$$V_{S''_N} \overline{A} = \begin{bmatrix} A_z \\ A_{nz} \end{bmatrix}.$$

Equation (9.2) can be rewritten as follows:

$$J_N(\chi,\delta) = \min_u \left\| \begin{bmatrix} A_z \\ A_{nz} \end{bmatrix} \begin{bmatrix} \chi \\ \delta \end{bmatrix} + \begin{bmatrix} 0 \\ B_{nz} \end{bmatrix} u \right\|_\infty$$

Thus, u cannot reduce the ℓ_∞ norm of $A_z \begin{bmatrix} \chi \\ \delta \end{bmatrix}$. If, for some χ and δ, $J_N(\chi,\delta) = \left\| A_z \begin{bmatrix} \chi \\ \delta \end{bmatrix} \right\|_\infty$, then the set of optimal solutions is given by

$$\{u \mid \left\| A_{nz} \begin{bmatrix} \chi \\ \delta \end{bmatrix} + B_{nz}u \right\|_\infty \leq \left\| A_z \begin{bmatrix} \chi \\ \delta \end{bmatrix} \right\|_\infty \},$$

which is not a singleton in general. Note that, if the simplex method is used to solve the linear program relative to Problem (9.2), then we obtain the optimal basic solution that also minimizes

$$\left\| A_{nz} \begin{bmatrix} \chi \\ \delta \end{bmatrix} + B_{nz} u \right\|_{\infty} .$$

However, it is computationally more efficient to directly compute the optimal u from

$$u^o = arg \min_{u} \left\| A_{nz} \begin{bmatrix} \chi \\ \delta \end{bmatrix} + B_{nz} u \right\|_{\infty} . \tag{9.3}$$

In this case, we need not to include the constraints from $\left\| A_z \begin{bmatrix} \chi \\ \delta \end{bmatrix} \right\|_{\infty}$ in the linear program.

We conclude the presentation of the optimality problem with a few comments. The result of Theorem 9.3.1 can be easily generalized to the case where the exogenous input is a vector rather than a scalar. However, a few remarks are in order.

Remark 9.3.1. In the scalar case, the only two possible worst-case strategies of the disturbance are $w = \delta$ and $w = -\delta$. For these values of the input, the worst-case perturbation of the state of the system is $B_1\delta$ and $-B_1\delta$. We removed the direct dependence on w by adding n_s new constraints.

If B_1 has n_w columns, then the disturbance has 2^{n_w} possible worst-case strategies to pick from. Thus, following the approach of Theorem 9.3.1, we would need to add $n_s(2^{n_w} - 1)$ new constraints in order to remove the dependence on w in the optimization problem. Therefore, Theorem 9.3.1 is of practical use only if n_w is small.

We need to consider approaches that reduce the complexity when n_w is not small. The approach we present here is standard in the $\mathcal{H}_\infty$ literature and has also been used in [18]. Instead of solving Problem (9.1) directly, we solve a feasibility problem, and then we use a bisection algorithm to approximate the optimal cost of Problem (9.1) with the desired accuracy.

9.3.3 Feasibility Problem

In this case, we are not interested in finding the optimal cost of Problem (9.1), rather, we want to check if a certain level of performance γ, for a given δ, is achievable. More precisely, we want to check if, starting from zero initial condition, $x_0 = 0$,

$$J_N(0,\delta) = \inf_{u_{[0:N-1]} \in R^{n_u \times N}} \max_{\|w_{[0:N-1]}\|_\infty \leq \delta} \max \left\{ \|z_{[0:N-1]}\|_\infty , J_0(x_N) \right\} < \gamma. \tag{9.4}$$

If the inequality holds, we say that γ is feasible.

From the previous section, it follows that $J_N(0,\delta)$ scales with δ, i.e., $J_N(0,\alpha\delta) = \alpha J_N(0,\delta)$ for any $\alpha \geq 0$. Thus, checking if $J_N(0,\delta) < \gamma$ is equivalent to check if $J(0,\frac{\delta}{\gamma}) < 1$. To simplify the notation, we define $\delta' = \frac{\delta}{\gamma}$, and let

$$H_N(\chi) = J_N(\chi, \delta').$$

Then, we have the following theorem. It says that if γ is feasible then $H_N(\chi)$ is a polytopic Gauge function that can be recursively computed.

Theorem 9.3.2. *Let Q_0 be a compact polytope in R^{n_s}, with n_s being the dimension of the system state space. Let R_0 be the unit ℓ_∞-ball in R^{n_z}. Consider the following recursion:*

$$Q^o_{k+1} = \tilde{A}^T \left\{ p \in P^o_k \mid p^T \tilde{B} = 0 \right\}$$

with

$$\tilde{A} = \begin{bmatrix} C_1 \\ A \end{bmatrix}, \qquad \tilde{B} = \begin{bmatrix} D_{12} \\ B_2 \end{bmatrix},$$

$$P^o_k = \begin{bmatrix} V^T_{R^o_0} & 0 \\ 0 & V^T_{S^o_k} \end{bmatrix} \mathcal{B}^o,$$

where $\mathcal{B}^o$ is the unit diamond of appropriate dimension, and

$$S^o_k = conv \left\{ s_i = \frac{q_i^T}{1 - \|q_i^T B_1\|_1 \delta'}, \mid q_i \in vertQ^o_k, \text{ and } \|q_i^T B_1\|_1 \delta' < 1 \right\}.$$

Then, γ in (9.4) is feasible for some integer N, if and only if

$$\|q_i^T B_1\|_1 \delta' < 1 \text{ for all } q_i \in vertQ^o_l \text{ and all } l = 0, \ldots, N-1,$$

and

$$H_N(\chi) = \max_{q \in Q^o_N} q^T \chi < 1.$$

Proof. See Appendix. ■

Notice that if γ is not feasible, then the algorithm has to be restarted with a greater γ. If γ is feasible, then a smaller feasible γ exists. Thus, if we want to get arbitrarily close to the optimal γ_0, several trial and error steps have to be performed, and this is efficiently done by implementing a bisection algorithm.

Relationship Between Optimality and Feasibility Problem. We would like to point out the close relationship between the two problems. Denote with $Q^o_{k,feas}$ the sets Q^o_k of the feasibility problem in Theorem 9.3.2, and denote with $Q^o_{k,opt}$ the sets Q^o_k of the optimality problem in Theorem 9.3.2. For a fixed $\delta \geq 0$ define

$$Q_{k,opt}(\delta) \triangleq \{x \in R^{n_s} \mid (x, \delta) \in Q_{k,opt}\},$$

notice that $Q_{k,opt}(\delta)$ can be empty.

For any $k \geq 1$, $Q_{k,opt}$ can be reinterpreted according to the following theorem:

Theorem 9.3.3. *Let $\Delta = \{\delta \geq 0 \mid Q_{N,opt}(\delta)$ not empty $\}$. Then, for any $\delta' \in \Delta$,*

$$Q_{k,opt}(\delta')^o = Q^o_{k,feas}$$

or, equivalently,

$$Q_{k,opt}(\delta') = Q^{oo}_{k,feas}$$

In other words, if we slice $Q_{k,opt}$ for some $\delta' \geq 0$, and the resulting set $Q_{k,opt}(\delta)$ is not empty, then it is equal to the closure of the set $Q_{k,feas}$.

Proof. Left to the reader. ∎

9.4 Infinite Horizon Full State Feedback

In this section we study the infinite horizon version of the optimality and feasibility problems. The following assumption holds throughout the section.

Assumption 9.4.1.

A1) C_1 and B_1 have rank n_s.
A2) $D_{12}^T C_1 = 0$.
A3) $Q_0 = \{x \mid \exists u$ such that $\|C_1 x + D_{12} u\|_\infty \leq 1\}$
A4) B_2 and D_{12} have full column rank.

We mainly concentrate on the optimality problem, since the arguments for the feasibility problem are similar. As shown in the next lemma, assumptions $A1 - A3$ guarantee that Q_k are compact and nested, i.e., $Q_{k+1} \subset Q_k$ for $k \geq 1$. The assumption that B_1 has rank n_s ensures that certain optimal and suboptimal strategies are globally exponentially stable. Assumption $A4$ implies the boundedness of the control input.

Lemma 9.4.1. *Under assumptions $A1 - A3$ the sets Q_k in Theorem 9.3.1 are nested, $Q_{k+1} \subset Q_k$, and compact for all $k \geq 1$.*

Proof. See Appendix. ∎

The nesting property implies that

$$Q_k = \cap_{i=1}^k Q_i \quad \text{for any } k \geq 1.$$

Since, for each $k \geq 1$, Q_k is nonempty (it contains $(0,0)$) and compact it follows from a well known analysis result [54] that

$$Q_\infty \triangleq \lim_{k \to \infty} Q_k = \cap_{i=1}^\infty Q_i$$

is nonempty and compact.

Next, we define the infinite horizon problem for a given initial state $x_0 = x$ and $\delta \geq 0$ as follows:

$$J^*(x,\delta) = \inf_u \max_{\|w\|_\infty \leq \delta} \|z\|_\infty \tag{9.5}$$

where u ranges over the stabilizing stationary full state feedback strategies $u_k = g(x_k)$.

We will show that $J^*(x,\delta)$ is actually given by the limit of the finite horizon problem as the length of the horizon goes to infinity.

Given the way Q_0 has been selected, each finite horizon problem of the form

$$J_N(x,\delta) = \inf_{u_{[0:N-1]} \in R^{n_u \times N}} \max_{\|w_{[0:N-1]}\|_\infty \leq \delta} \max\left\{\|z_{[0:N-1]}\|_\infty, J_0(x_N)\right\}$$

is equivalent to the following problem: given $x_0 = x$ and $\delta \geq 0$

$$J_N(x,\delta) = \inf_{u_{[0:N]} \in R^{n_u \times N+1}} \max_{\|w_{[0:N]}\|_\infty \leq \delta} \|z_{[0:N]}\|_\infty \tag{9.6}$$

For any (x,δ) we can define $J_\infty(x,\delta)$ by taking the limit as N goes to infinity of $J_N(x,\delta)$.

$$J_\infty(x,\delta) = \lim_{N\to\infty} J_N(x,\delta) = \lim_{N\to\infty} \max_{q \in Q_N^o} q^T \begin{pmatrix} x \\ \delta \end{pmatrix} = \max_{q \in Q_\infty^o} q^T \begin{pmatrix} x \\ \delta \end{pmatrix}.$$

Note that $J_\infty(x,\delta)$ can, in principle, be unbounded. Next result states that this cannot happen.

Lemma 9.4.2. *For any x and $\delta \geq 0$*

$$J_\infty(x,\delta) \leq J^*(x,\delta).$$

Proof. Given $x = x_0$ and $\delta \geq 0$, $J^*(x,\delta)$ is bounded since any internally stabilizing linear controller will ensure BIBO stability. Since $Q_{k+1} \subset Q_k$, then $J_{k+1}(x,\delta) \geq J_k(x,\delta)$ for any x and $\delta \geq 0$. The sequence $\{J_N(x,\delta)\}$ is therefore monotonically non decreasing. Given any $\epsilon > 0$, there is a stabilizing stationary full state feedback strategy that achieves performance $J^*(x,\delta) + \epsilon$ on the infinite horizon. The same strategy is certainly suboptimal for the finite horizon problem. Then $J_N(x,\delta)$ is bounded above by $J^*(x,\delta) + \epsilon$. Thus $\{J_N(x,\delta)\}$ is convergent and this implies that

$$J_\infty(x,\delta) \leq J^*(x,\delta).$$

■

In the next results, we concentrate our attention to $J_\infty(x,\delta)$. We first show that, for a given $x \in R^{n_x}$ and $\delta > 0$, there is a stationary full state feedback strategy achieving $J_\infty(x,\delta)$.

Lemma 9.4.3. *For any $x \in R^{n_s}$ and any $\delta > 0$ there exists a stationary full state feedback strategy that achieves the performance $J_\infty(x, \delta)$.*

Proof. From the definition of J_∞ it follows that J_∞ satisfies the following Bellman equation.

$$J_\infty(x,\delta) = \inf_u \max_{\|w\|_\infty \leq \delta} \max\{\|C_1 x + D_{12}u\|_\infty, J_\infty(Ax + B_2 u + B_1 w, \delta)\}$$

For a given $\gamma \geq 0$, $\delta \in R^+$, and $x \in R^{n_s}$, the set $\{u \mid J_\infty(x,\delta) \leq \gamma\}$ is either empty or is included in a compact set $\{u \mid \|u\| \leq \beta\}$ for some $\beta < \infty$. This is because, by Assumption *A4*, $\|C_1 x + D_{12}u\|_\infty$ will go unbounded if $\|u\| \to \infty$. Thus, $J_\infty(x, \delta)$ is equivalently given by

$$J_\infty(x,\delta) = \inf_{\|u\| \leq \beta} \max_{\|w\|_\infty \leq \delta} \max\{\|C_1 x + D_{12}u\|_\infty, J_\infty(Ax + B_2 u + B_1 w, \delta)\}$$

for a large enough but bounded β. Since we are minimizing a convex function of u over a compact set, there exists a bounded u for which the infimum is actually achieved. ∎

Notice that the above lemma is proving that, whenever $Q_\infty(\delta)$ is not empty, it is controlled invariant according to Definition 9.4.1 given later and taken from [19].

Corollary 9.4.1. *Q_∞ has a nonempty interior.*

From these results it follows that, for $\delta > 0$, $J^*(x, \delta) > 0$ for all $x \in R^{n_s}$. Thus, under the current assumptions the optimal cost is never zero.

Note that, if $Q_\infty(\delta)$ is nonempty for some $\delta > 0$, then it has a nonempty interior in R^{n_s}. This is true since $Q_\infty(\delta)$ is controlled invariant, and therefore any optimal strategy takes any x on the boundary of $Q_\infty(\delta)$ into an element of the set:

$$\mathcal{S}_\infty(\delta) = \{\xi \mid \xi + B_1 w \in Q_\infty(\delta),\ \forall w : \|w\|_\infty \leq \delta\}$$

Now, $\mathcal{S}_\infty(\delta)$ is strictly contained in $Q_\infty(\delta)$ (since B_1 has rank n_s by Assumption *A1*), and it contains the origin (x=0).

From this property and the compactness of Q_∞, it follows that

$$\delta^o = \max\{\delta > 0 \mid Q_\infty(\delta) \neq \emptyset\}$$

is well defined and $Q_\infty(\delta^o)$ has nonempty interior.

Consider now, $J_\infty(0, \delta^o)$. Form the above discussion it follows that $J_\infty(0, \delta^o) = 1$. Denote with $u^o = \tilde{g}(x)$ a strategy that achieves $J_\infty(0, \delta^o)$ (in Lemma 9.4.3). As already pointed out, $\tilde{g}(x)$ maps elements on the boundary of $Q_\infty(\delta^o)$ into elements of $\mathcal{S}_\infty(\delta^o) \subset Q_\infty(\delta^o)$. This strategy may not be homogeneous in x, i.e., $\tilde{g}(\alpha x) \neq \alpha \tilde{g}(x)$ for $\alpha \geq 0$; however, a homogeneous strategy can be derived by appropriate scaling of $\tilde{g}$ following the idea in [18]. Let $\rho(x)$, the scaling factor, be defined as follows:

$$\rho(x) \triangleq \inf_{x \in \rho Q_\infty(\delta^o)} \rho = \max_{q \in Q''_\infty(\delta^o)} q^T x$$

Now define the new strategy $g : R^{n_x} \to R^{n_u}$ as follows: Set $g(0) = 0$, and

$$g(x) = \rho(x)\tilde{g}(x/\rho(x)). \tag{9.7}$$

Then, it is easy to verify that g has the desired scaling property: $g(\alpha x) = \alpha g(x)$ for $\alpha \geq 0$ and it is equal to $\tilde{g}(x)$ on the boundary of $Q_\infty(\delta^o)$.

We next show that g is a globally exponentially stable strategy and achieves the performance $\gamma^o = 1/\delta^0$ uniformly, i.e., independently of the maximum input amplitude.

Lemma 9.4.4. *The strategy defined in Equation (9.7) is globally exponentially stable and guarantees an ℓ_∞-induced gain equal to $1/\delta^o = \gamma^o$ independently of the input magnitude bound.*

Proof. Follows immediately from Claim 3.10 and Claim 3.11 in [18]. Here we just sketch it. We can use the gauge function defined by $Q_\infty(\delta^o)$ as a norm on R^{n_x}:

$$\|x\| = \min_{x \in \alpha Q_\infty(\delta^o)} \alpha.$$

Since $\mathcal{S}_\infty(\delta^o)$ is strictly included in $Q_\infty(\delta^o)$ we have that

$$\beta = \max_{x \in \mathcal{S}_\infty(\delta^o)} \|x\| < 1$$

By the scaling property it follows that, under the strategy $g(x)$, with $w = 0$, any element $x_k \in R^{n_x}$ is brought to an element x_{k+1} such that $\|x_{k+1}\| \leq \beta \|x_k\|$. Thus

$$\|x_k\| \leq \beta^{k+1} \|x_0\|.$$

The optimality of the strategy follows from the fact that $\tilde{g}$ is optimal and g coincides with $\tilde{g}$ on the boundary of $Q_\infty(\delta^o)$. ■

We combine the results of Lemma 9.4.2 Lemma 9.4.3 and Lemma 9.4.4 in the next theorem.

Theorem 9.4.1. *The strategy defined in Equation (9.7) is an optimal infinite horizon strategy, i.e.,*

$$J^*(0, \delta) = \gamma^o \delta$$

Proof. From Lemma 9.4.2 we know that $J_\infty(x, \delta) \leq J^*(x, \delta)$. Form Lemma 9.4.3 and Lemma 9.4.4 we have that the strategy Equation (9.7) is stationary globally stabilizing and achieves performance γ^o independently of the maximum input amplitude δ. The result then follows immediately. ■

Although in some cases, as for example when the recursion converges to Q_∞ in a finite number of steps, Q_∞ is a polytope, in general, Q_∞ is a well defined convex set, but it may not be a polytope. Even if it is a polytope, it may have too many vertices to be computed in a reasonable time. Therefore, we need ways to compute stabilizing suboptimal strategies.

9.4.1 Stabilizing Suboptimal Strategies

In this section, we show how to derive suboptimal strategies that result in closed loop performance arbitrarily close to the optimal one and closed loop system exponentially stable. We present two ways of deriving such suboptimal strategies.

The first one is based on computing an upper bound on the achievable performance when, at each step, we use the strategy that is optimal at time 0 for the finite horizon problem scaled similarly to Equation(9.7). As the length of the horizon increases, the optimal strategy at time zero will look more and more like the stationary optimal strategy for the infinite horizon.

The second approach is basically the method proposed in [19]. As shown in Section 9.4.1, there is a practical equivalence between the feasibility problem in Section 9.3.3 and the invariant kernel algorithm used in [19]. Thus, due to the relation between the optimality problem and the feasibility problem, highlighted in Section 9.3.3, the results given in [19] can be applied to obtain suboptimal strategies.

Method #1. In order to properly describe this method, we need to introduce some notation and derive some preliminary results.

Consider the finite horizon optimality problem, with horizon of length N. As discussed in Section 9.3.2, the optimal strategy is generally different for each step of the horizon. Consider the optimal strategy at step 0 (assumed unique for simplicity) given by Equation (9.3). Fix a $\delta \geq 0$ such that $Q_N(\delta)$ is nonempty, and denote with $\mathcal{S}_{N-1}(\delta)$ the following subset of $Q_{N-1}(\delta)$

$$\mathcal{S}_{N-1}(\delta) = \{\xi \in R^{n_x} \mid \xi + B_1 w \in Q_{N-1}(\delta), \quad \forall w : \|w\|_\infty \leq \delta\},$$

and denote with $\tilde{g}(x)$ the optimal strategy at step 0, $u^o = \tilde{g}(x)$.

From the definition of $Q_N(\delta)$ it follows that, through the optimal strategy, each element on the boundary of $Q_N(\delta)$ is mapped into an element of $\mathcal{S}_{N-1}(\delta)$. More formally, if we denote with $BQ_N(\delta)$ the boundary of $Q_N(\delta)$, and with $\mathcal{R}_N(\delta)$ the set of reachable states from $BQ_N(\delta)$ in one step under the optimal strategy at step 0, namely

$$\mathcal{R}_N(\delta) = \{y = Ax + B_2\tilde{g}(x) \mid x \in BQ_N(\delta)\},$$

then,

$$\mathcal{R}_N(\delta) \subset \mathcal{S}_{N-1}(\delta).$$

Moreover, from the property of the optimal solution, it follows that the intersection of the boundary of $\mathcal{R}_N(\delta)$ and the boundary of $\mathcal{S}_{N-1}(\delta)$ is not empty.

Note that the following is true:

Lemma 9.4.5. *If, for some $\delta \geq 0$, $Q_N(\delta)$ has nonempty interior, and $\mathcal{S}_{N-1}(\delta)$ is strictly contained in the interior of $Q_N(\delta)$, i.e., if there exists a $\alpha > 0$ such that, for any $y \in \mathcal{S}_{N-1}(\delta)$,*

$$y + v \in Q_N(\delta) \qquad \forall\, v : |||v||_\infty \leq \alpha,$$

then, there exists a $\delta' \leq \delta$ such that, for any $x \in Q_N(\delta)$, under the optimal strategy at step 0,

$$Ax + B_2\tilde{g}(x) + B_1 w \in Q_N(\delta),\ \forall\, w : \|w\|_\infty \leq \delta'$$

In other words, $Q_N(\delta)$ is controller invariant under the optimal strategy at step 0, when $\|w\|_\infty \leq \delta'$.

Proof. Under the optimal strategy at step 0

$$y = Ax + B_2\tilde{g}(x) \in \mathcal{R}_N(\delta) \subset \mathcal{S}_{N-1}(\delta), \qquad \forall,\, x \in Q_N(\delta)$$

Let $v = B_1 w$, then for $\delta' \leq \frac{\alpha}{\|B1\|_1}$, we have that $\|v\|_\infty \leq \alpha$ for all w with $\|w\|_\infty \leq \delta'$. Thus for any $x \in Q_N(\delta)$ we have that

$$Ax + B_2\tilde{g}(x) + B_1 w \in Q_N(\delta),\ \forall\, w : \|w\|_\infty \leq \delta'$$

■

Note that, if the assumptions of the above lemma are satisfied, then, the optimal strategy at step 0 will achieve a performance level, on the infinite horizon, less than or equal to 1 for all initial conditions $x \in Q_N(\delta)$, when $\|w\|_\infty \leq \delta'$.

We can now derive a homogeneous strategy from the optimal strategy at step 0, so that the new strategy is globally exponentially stable and achieves the performance level $1/\delta'$ independently of the actual input magnitude bound.

We follow the same procedure used for the optimal infinite horizon strategy. Let $\rho(x)$, the scaling factor, be defined as follows

$$\rho(x) \triangleq \inf_{x \in \rho Q_N(\delta)} \rho = \max_{q \in Q_N^o(\delta)} q^T x$$

We define the new strategy $g : R^{n_x} \to R^{n_u}$ as follows: Set $g(0) = 0$, and

$$g(x) = \rho(x)\tilde{g}(x/\rho(x)) \tag{9.8}$$

g has the following scaling property: $g(\alpha x) = \alpha g(x)$ for $\alpha \geq 0$.

Following the development in Lemma 9.4.4, thus, we obtain the following result:

Lemma 9.4.6. *Under the assumptions of Lemma 9.4.5, the strategy defined in Equation (9.8) is globally exponentially stable and guarantees an ℓ_∞-induced gain less than or equal to $1/\delta'$, independently of the input magnitude bound.*

Proof. Follows immediately from Claim 3.10 and Claim 3.11 in [18]. ∎

The approach we adopt is now clear. We first derive the optimal strategy at step 0 and then, by using Equation (9.8), we transform it into an exponentially stabilizing strategy. We only need to ensure that the assumptions of Lemma 9.4.5 are satisfied. However, the applicability of Lemma 9.4.5 is guaranteed under the current assumptions ($A1 - A4$). We know that $\{Q_N\}$ is a sequence of compact sets converging to a nonempty compact set Q_∞ (Lemma 9.4.1). Thus, for N large enough, Q_N and Q_{N-1} will be arbitrarily close in the Hausdorff metric. The assumption that B_1 has rank n_s, then, guarantees that, for $\delta > 0$, $Q_{N-1}(\delta)$ strictly includes in its interior $\mathcal{S}_{N-1}(\delta)$. Then, there exists a $\delta > 0$ such that $Q_N(\delta)$ is nonempty and strictly includes in its interior $\mathcal{S}_{N-1}(\delta)$. Summarizing, we have

Theorem 9.4.2. *Under the current assumptions, there exist $N \geq 2$, $\delta \geq 0$, $0 \leq \delta' \leq \delta$, and a globally exponentially stable strategy constructed accordingly to Equation (9.8), which guarantees an ℓ_∞-induced gain $\gamma = 1/\delta'$ for the infinite horizon problem, independent of the input magnitude amplitude.*

Proof. The result follows from Lemma 9.4.5 and Lemma 9.4.6. ∎

In particular, for any $\delta \in (0, \delta^o]$ we have the following stronger result:

Corollary 9.4.2. *Given any $\epsilon > 0$ and any $\delta \in (0, \delta^o]$, there exist a $\delta' \in [\delta - \epsilon, \delta]$ and a N large enough such that a globally exponentially stable strategy for the infinite horizon problem can be constructed accordingly to Equation (9.8) with guaranteed ℓ_∞-induced gain*

$$\gamma = 1/\delta'$$

independent of the input amplitude.

Proof. The fact that $\delta \in (0, \delta^o]$ implies that $Q_{N-1}(\delta)$ and $Q_N(\delta)$ are nonempty for any N and approaching each other as $N \to \infty$. At the same time, since B_1 has rank n_s, there is a $\beta > 0$ independent of N such that, for any $x \in \mathcal{S}_{N-1}(\delta)$, $x + v \in Q_{N-1}(\delta)$ for all v with $\|v\| \leq \beta$. ∎

The result of this corollary is that a cost arbitrarily close to the optimal one, $\gamma^o = 1/\delta^o$, can be achieved by the scaled version of the optimal finite horizon strategy at step 0 as the length of the horizon increases.

Remark 9.4.1. The above result shows that J^* defined in Equation (9.5) is the optimal cost over exponentially stabilizing strategies.

In practice, the length of the horizon we can consider is limited by the complexity of computing the vertices of Q_N. For a given N and δ we can check if $\mathcal{S}_{N-1}(\delta)$ is strictly included in the interior of $Q_N(\delta)$. If it is, we know by Lemma 9.4.5 and Lemma 9.4.6 that the scaled optimal finite horizon strategy at step 0 will be globally exponentially stabilizing and achieve a performance $1/\delta'$ for some $\delta' \leq \delta$.

Note that the condition that $\mathcal{S}_{N-1}(\delta)$ is strictly contained in the interior of $\mathcal{Q}_N(\delta)$ can be easily checked by solving a sequence of linear programs involving the polars of the two sets. In practice, we only need to compute δ' or a lower bound for it. This is accomplished by finding the largest $\delta' > 0$ that guarantees the following inclusion

$$\mathcal{S}_N(\delta') \supset \mathcal{S}_{N-1}(\delta). \tag{9.9}$$

Clearly, any $\delta' > 0$ with the above property is a lower bound for the maximum input amplitude consistent with the invariance of $\mathcal{Q}_N(\delta)$. Thus, $1/\delta'$ is an upper bound on the achievable performance when we use the strategy g, which, in turn, is an upper bound to the optimal infinite horizon cost.

The set inclusion in (9.9) can be tested as follows. For fixed δ and $\delta' < \delta$, if

$$\mu = \max_{s_i \in vert\mathcal{S}_N^o(\delta')} \max_{V_{\mathcal{S}_{N-1}^o(\delta)}\, x \leq 1} s_i^T x \leq 1,$$

then (9.9) is satisfied.

The sets $\mathcal{S}_N(\delta')$ and $\mathcal{S}_{N-1}(\delta)$ are relatively easy to compute. $\mathcal{S}_N^o(\delta')$ can be computed from $\mathcal{Q}_N(\delta)$ as follows:

$$\mathcal{S}_N^o = conv\left\{ s_i = \frac{q_i^T}{1 - \|q_i^T B_1\|_1 \delta'},\ |\ q_i \in vert\mathcal{Q}_k^o(\delta) \right\}$$

An analogous relation holds for $\mathcal{S}_{N-1}^o(\delta)$.

Finally, given $\mathcal{Q}_k^o$ and $\delta \geq 0$ (strictly smaller than the largest δ^o for which $\mathcal{Q}_k(\delta^o)$ is not empty), $\mathcal{Q}_k^o(\delta)$ is given by

$$\mathcal{Q}_k^o(\delta) = \left\{ x \in R^{n_s} \mid [q_1^T, q_2^T] \begin{bmatrix} x \\ \delta \end{bmatrix} \leq 1, \forall\, q \in \mathcal{Q}_k^o, \text{ with } q^T = [q_1^T, q_2^T] \right\}^o$$

or, equivalently,

$$\mathcal{Q}_k^o(\delta) = conv\left\{ v_i = \frac{q_{1i}}{1 - q_{2i}^T \delta} \mid \quad \forall\, [q_{1i}, q_{2i}]^T \in vert\mathcal{Q}_k^o \right\}.$$

Before we describe the next method, we investigate the relation between the feasibility problem of Section 9.3.3 and the invariant kernel algorithm used in [19].

Feasibility Problem and Invariant Kernel Algorithm. In this section, we will see how the result of Theorem 9.3.2 regarding the feasibility problem compares with the method in [19]. For simplicity of exposition, we further assume that $D_{12} = 0$. In this and the next section, $\mathcal{Q}_k$ and $\mathcal{S}_k$ denote the sets of the feasibility problem defined in Section 9.3.3. In this section, we establish that

- The sets K_j and $K_{j\frac{1}{2}}$ in [19] (see later) are the sets Q_k and S_k, for a particular choice of γ and δ in our notation.
- In [19], there is no need to introduce the Rack operator, and all the desired sets can be derived from standard properties of polar sets.
- The dynamic programming approach reduces the complexity in the computation of Q_k^o at each step.

To avoid confusion between the different notations, we denote with γ_s the γ in [19]. Let $\gamma = 1$ and $\delta = 1/\gamma_s$ in the setup of this chapter. We now briefly recall the notation and the results in [19]. For simplicity, we describe the case of scalar control under the current hypothesis.

Controlled Invariant Kernel Algorithm
Consider the controlled difference inclusion $(F_{\gamma_s}, U_{\gamma_s}, f)$ where

$$\begin{aligned} F_{\gamma_s}(\xi) &= \left\{\xi + \frac{1}{\gamma_s} B_1 w \in R^{n_s} \mid |w| \le 1\right\} \\ U_{\gamma_s}(x) &= \{u \in R \mid |C_1 x| \le 1\} \\ f(x,u) &= Ax + B_2 u \end{aligned}$$

Let $dom(U_{\gamma_s}) = \{x \in R^{n_s} \mid |C_1 x| \le 1\}$.

Definition 9.4.1. *A subset $K \subset dom(U_{\gamma_s})$ is controlled invariant under the difference inclusion, $(F_{\gamma_s}, U_{\gamma_s}, f)$, if for every $x \in K$, there exists a $u \in U_{\gamma_s}(x)$ such that $F(f(x,u)) \subset K$.*

Definition 9.4.2. *The largest closed subset of K which is controlled invariant under the difference inclusion $(F_{\gamma_s}, U_{\gamma_s}, f)$ is the controlled invariant kernel of K and is denoted by $C_{INV}(K)$.*

Let $K \subset dom(U_{\gamma_s})$ be compact. Define recursively the subsets K_j and $K_{j\frac{1}{2}}$ by

$$\begin{aligned} K_0 &= K \\ K_{j\frac{1}{2}} &= \{\xi \in dom(F_{\gamma_s}) \mid F(\xi) \subset K_j\} \\ K_{j+1} &= \left\{x \in K_j \mid f(x,u) \in K_{j\frac{1}{2}} \text{ for some } u \in U_{\gamma_s}(x)\right\} \end{aligned} \tag{9.10}$$

Then

$$C_{INV}(K) = \bigcap_{j=0}^{\infty} K_j.$$

For $M \in R^{z,n_s}$, Set(M) denotes the subset of R^{n_s} associate with M defined by the constraints:

$$Set(M) = \{x \in R^{n_s} \mid Mx \le 1\}$$

Notice that, in our notation $M = V_{(Set(M))^o}$.

In [19], the Rack operator was introduced to characterize the solution to the following problem: Given $M = [M_1, M_2]$, with $M_1 \in R^{l \times k}$ and $M_2 \in R^{l \times 1}$, and the set

$$\mathcal{S} = \{(v,w) \in R^k \times R \,|\, M_1 v + M_2 w \le 1\},$$

find the set $\tilde{\mathcal{S}} = \{v \in R^k \,|\, (v,w) \in \mathcal{S}$ for some $w \in R\}$. $Rack[M]$ is the set of all matrices, $\overline{M}$, that describe the set $\tilde{\mathcal{S}}$ as follows:

$$\tilde{\mathcal{S}} = \{v \in R^k \,|\, \overline{M} v \le 1\}$$

Theorem 9.4.3. *Let $M = (M_1, M_2)$, where M_1 has n columns and M_2 has one column. Let*

$$\begin{aligned} Z_+ &= \{i \,|\, (M_2)_i > 0\} \\ Z_- &= \{i \,|\, (M_2)_i < 0\} \\ Z_0 &= \{i \,|\, (M_2)_i = 0\} \end{aligned}$$

Let the matrix $\tilde{M}$ be formed by the rows

$$\rho^T_{i_+i_-} = \frac{1}{(M_2)_{i_+} - (M_2)_{i_-}}((M_2)_{i_+}(M_1)_{(i_-,:)} - (M_2)_{i_-}(M_1)_{(i_+,:)}), \quad \forall\, i_+ \in Z_+\; i_- \in Z_- \tag{9.11}$$

$$\rho^T_{i_0} = (M_1)_{i_0,:} \quad \forall i_0 \in Z_0$$

Then $\tilde{M} \in Rack[M]$

The algorithm implementation is given by the following steps. Assume that $\gamma_s > \gamma^o$.

1) Initialize $K_0 = dom(U_{\gamma_s})$. Define $M_0 = \begin{pmatrix} C_1 \\ -C_1 \end{pmatrix}$. Note that $Set(M_0) = K_0$ and in our notation $M_0 = V_{K_0^o}$.

2) Then $K_{j\frac{1}{2}} = Set(M_{\frac{1}{2}})$ where

$$(M_{j\frac{1}{2}})_{(i,:)} = \frac{1}{1 - \frac{1}{\gamma_s}|(M_j B_1)_{(i,:)}|}(M_j)(i,:).$$

3) Set

$$M_{j+1} = \begin{pmatrix} M_j \\ N \end{pmatrix} \tag{9.12}$$

for any

$$N \in Rack\left[M_{j\frac{1}{2}}A,\; M_{j\frac{1}{2}}B_2\right]. \tag{9.13}$$

Then

$$C_{INV}(dom(U_{\gamma_s})) = \bigcap_{j=0}^{\infty} Set(M_j)$$

Comparison with the Feasibility Problem

From Assumption $A3$ we have that

$$Q_0 = \left\{ x \mid \begin{bmatrix} C_1 \\ -C_1 \end{bmatrix} x \leq 1 \right\}, \tag{9.14}$$

then, the sets Q_0 and S_0 are nothing but the sets K_0 and $K_{\frac{1}{2}}$ in (9.10). Also Q_0^o and S_0^o correspond to M_0 and $conv\{\text{rows of } M_{\frac{1}{2}}\}$ respectively. In particular, we have that $V_{Q_0^o} = M_0$ and $V_{S_0^o} = M_{\frac{1}{2}}$. Also, from the definition of the set Q_1 we have that

$$Q_1 = \{x \in R^{n_s} \mid \exists u \,\text{such that}\; \|C_1 x\|_\infty \leq 1, \; and \\ Ax + B_2 u + B_1 w \in Q_0, \forall w : \|w\|_\infty \leq \delta\}.$$

However, from Equation (9.14) and the definition of S_0, Q_1 can be rewritten as:

$$Q_1 = \{x \in Q_0 \mid Ax + B_2 u \in S_0 \;\; \text{for some } u\},$$

but, given that $Q_0 = K_0$ and $S_0 = K_{\frac{1}{2}}$, this is the definition of K_1 in (9.10). The equality between Q_j and K_j for $j > 1$ follows from a similar argument and the property of Q_j of being nested (Lemma 9.4.1).

For our purposes, the characterization of $Rack[M]$ is not necessary. All we need is to find some matrix $\overline{M} \in Rack[M]$. In order to do so, we only need to characterize $\tilde{S}$ in terms of $\tilde{S}^o$. Toward this end, define $\mathcal{M} = \{z = M_2 w, \mid w \in R\}$, and consider the following set:

$$\overline{S} = \{\xi \in R^l \mid \xi + M_2 w \in \mathcal{B}, \;\; \text{for some } w \in R\}.$$

Then $\overline{S} = \mathcal{B} + \mathcal{M}$, where + means set addition. From the properties of polar sets [53], it follows that the polar of $\overline{S}$ is given by the following expression:

$$\overline{S}^o = \mathcal{B}^o \cap \mathcal{M}^\perp,$$

with $\mathcal{M}^\perp = \{s \in R^l \mid s^T M_2 = 0\}$. Given $\overline{S}^o$, the set $\tilde{S}$ can be expressed as follows:

$$\tilde{S} = \{v \in R^k \mid M_1 v \in \overline{S}\}.$$

Thus, from [53], the set $\tilde{S}^o$ is given by

$$\tilde{S}^o = M_1^T \, \overline{S}^o.$$

We show next that the matrix $V_{\tilde{S}^o}$ belongs to $Rack[M]$, and in particular is nothing but the matrix $\tilde{M}$ in [19]. At the same time, we see that the operator $Rack[M]$ does not play any role in this derivation, since the important matrix, $V_{\tilde{S}^o}$, is naturally related with the polar of the set $\tilde{S}$.

$\overline{S}^o$ is given by the intersection of the unit diamond in R^l, $\mathcal{B}^o$, with $\mathcal{M}$. The $2l$ vertices of $\mathcal{B}^o$ are $e_1, \ldots, e_l, -e_1, \ldots, -e_l$, where e_i is the ith basis vector of the Euclidean orthonormal basis of R^l. The sets Z_+, Z_-, and Z_0 in [19] are characterized by the inner product of M_2 with the vertices of $\mathcal{B}^o$. Thus,

$$Z_+ = \{i \mid e_i^T M_2 > 0\}$$
$$Z_- = \{i \mid e_i^T M_2 < 0\}$$
$$Z_0 = \{i \mid e_i^T M_2 = 0\}$$

The set of vertices of $\overline{S}^o$ is then given by

$$s_{ij} = \frac{1}{(M_2)_i - (M_2)_j}((M_2)_j e_i - (M_2)_i e_j), \quad \forall i \in Z_+, \forall j \in Z_-$$
$$\text{Union}$$
$$s_{i_0} = e_{i_0}, \quad \forall i_o \in Z_0$$

It is immediate to verify that $\rho_{i_+ i_-}$ and ρ_{i_0} in Equation 9.11. are the images of the vertices of $\overline{S}^o$ through M_1^T.

From this interpretation of $\tilde{M}$, we also see that the vertices s_{ij} with $|i - j| = l$ are internal vertices of $\overline{S}^o$ and, given the convexity of M_1^T, they result in internal vertices of $\tilde{S}^o$. Thus, they can be omitted. Notice, however, that other vertices may be internal to $\tilde{S}^o$. The external vertices of $\tilde{S}^o$ are given by $vert(conv(M_1^T vert \overline{S}^o))$. Thus, we have shown that $\tilde{S}$ is the set whose polar is $\tilde{S}^o = M_1(\mathcal{B}^o \cap \mathcal{M}^\perp)$.

The above discussion it is also useful to show the computational advantage of the dynamic programming approach with respect to [19]. Apply the result above to the set

$$\mathcal{S} = \{(x,u), \mid \begin{bmatrix} V_{R_0^o} C_1 \\ V_{S_0^o} A \end{bmatrix} x + \begin{bmatrix} V_{R_0^o} D_{12} \\ V_{S_0^o} B_2 \end{bmatrix} u \in \mathcal{B}\}.$$

From the previous derivation we have that $\tilde{S}^o$ is equal to

$$\tilde{\mathcal{S}}^o = \begin{bmatrix} V_{R_0^o} C_1 \\ V_{S_0^o} A \end{bmatrix}^T \left(\mathcal{B}^o \cap \left\{ s \mid s^T \begin{bmatrix} V_{R_0^o} D_{12} \\ V_{S_0^o} B_2 \end{bmatrix} = 0 \right\} \right)$$

Then, it follows that $\tilde{S}^o$ is nothing but the set Q_1^o defined in Theorem 9.3.2. From this and the fact that $V_{Q_0^o} = M_0$ and $V_{S_0^o} = M_{\frac{1}{2}}$, we deduce that

$$Q_1^o \in Rack \left[\begin{pmatrix} M_0 C_1 & M_0 D_{12} \\ M_{\frac{1}{2}} A & M_{\frac{1}{2}} B_2 \end{pmatrix} \right].$$

In [19], however, the matrix $V_{\tilde{S}^o}$ is denoted by N (given by Equation (9.13)) and, following step 3) of the implementation of the algorithm, K_1 is identified with the set

$$K_1 = \left\{ x \mid \begin{bmatrix} M_0 \\ N \end{bmatrix} x \le 1 \right\}$$

or

$$K_1 = set\left(\begin{bmatrix} M_0 \\ N \end{bmatrix} \right) = set(M_1)$$

in [19] notation. But, from our discussion, we have that

$$Q_1 = \{x \mid Nx \le 1\} = set(N).$$

Therefore, we immediately have that the vertices that define Q_1^o are the rows of N, while in Equation (9.12) the vertices defining Q_1^o are taken to be the rows of $M_1 = \begin{bmatrix} M_0 \\ N \end{bmatrix}$.

Thus, using dynamic programming, we are able to detect all the rows of M_0 in the definition of M_1, or, in general, all the rows of M_j in the definition of M_{j+1} that are redundant and, therefore, can be eliminated a priori. As already pointed out, not all the redundancy is removed, since

$$\begin{aligned} vertQ_j^o &= vert\left(conv\tilde{A}^T vert\left\{p \in P_{j-1}^o \,|\, p^T\tilde{B} = 0\right\}\right) \\ &\subset \tilde{A}^T vert\left\{p \in P_{j-1}^o \,|\, p^T\tilde{B} = 0\right\}. \end{aligned}$$

The above arguments prove that the invariant kernel algorithm is a dynamic programming algorithm for the feasibility problem for a particular choice of the terminal cost function.

Method #2. Having established that the recursion in Theorem 9.3.2 and the recursion presented in [19] are the same in terms of resulting sets, we can apply the same method developed in [19] to derive stabilizing suboptimal nonlinear controllers.

The following results from [19] are rephrased here for convenience.

Theorem 9.4.4. *If, for a given δ, γ is feasible, and the sequence of polyhedral sets converges in a finite number of steps, say $Q_N = Q_\infty$, then, one can construct a nonlinear internally stabilizing controller that will achieve, within any $\epsilon > 0$, the performance γ/δ.*

When Q_∞ cannot be achieved in a finite number of steps, then the following stopping criterion can be adopted.

Theorem 9.4.5. *Given Q_k and $\eta \geq 0$, define*

$$\begin{aligned} \tilde{Q}_{k+1} = \{x \in R^{n_x} \,|\, \exists u \text{ such that } &\|\tfrac{C_1}{\sqrt{1+\eta}}x + \tfrac{D_{12}}{\sqrt{1+\eta}}u\|_\infty \leq 1, \\ &\text{and } Ax + B_2u + \tfrac{B_1}{\sqrt{1+\eta}} \in Q_k,\ \forall w : \|w\|_\infty \leq \delta' = \delta/\gamma\ \} \end{aligned}$$

If, for a given δ, γ is feasible for the infinite horizon problem, then, for any $\eta > 0$, there is a N such that $Q_k \subset \tilde{Q}_{k+1}$ and one can construct a stabilizing controller that achieves, within any $\epsilon > 0$, a closed loop performance $\frac{\gamma}{\delta}(1+\eta)$.

The above theorem can be used to derive an upper bound to the optimal cost. If, for the given N, γ, and δ, one can find $\eta \geq 0$ for which $Q_N \subset \tilde{Q}_{N+1}$, then it is possible to construct a stabilizing controller that achieves a closed loop performance $\frac{\gamma}{\delta}(1+\eta)$. Note however that, for a given N, there may be no $\eta > 0$ that ensures the inclusion of Q_k by $\tilde{Q}_{k+1}$. In this cases, either N must be increased, or γ must be increased and the invariant kernel algorithm restarted.

When we consider the optimality problem, by Theorem 9.3.3, we can apply the above theorem to the sets $Q_N(\delta')$ and $\tilde{Q}_{N+1}(\delta')$, where $\tilde{Q}_{N+1}$ is defined as

$$\tilde{Q}_{N+1} = \{(x,\delta) \in R^{n_x} \times R \,|\, \exists u \text{ such that } \|\tfrac{C_1}{\sqrt{1+\eta}}x + \tfrac{D_{12}}{\sqrt{1+\eta}}u\|_\infty \leq 1,$$
$$\text{and } (Ax + B_2 u + \tfrac{B_1}{\sqrt{1+\eta}}, \delta) \in Q_N,\ \forall w : \|w\|_\infty \leq \delta\ \}$$

In this case, since we have already computed Q_N, we can compute $Q_N(\delta')$, for any small enough δ', relatively easily, without having to restart the recursion. Still, we may have to increase N in order to find a suitable η.

Comment: We conclude this section with a few comments. We want to point out that, if we are solving an optimality problem, Method #1 is computationally more efficient than Method #2. This is due to two factors: First, the inclusion (9.9) is more easily satisfied than the inclusion $Q_N(\delta) \subset \tilde{Q}_{N+1}(\delta)$, since $\tilde{Q}_{N+1}(\delta)$ is non trivially dependent on η. Second, in Method #2, if the inclusion $Q_N(\delta) \subset \tilde{Q}_{N+1}(\delta)$ is not satisfied, and we need to go a step further in the recursion, we need not only to compute Q_{N+1}, but also to recompute $\tilde{Q}_{N+2}$ from Q_{N+1}, while $\tilde{Q}_{N+1}$ goes wasted. In contrast, Method #1 only needs Q_{N+1} and Q_N, and the sets $\mathcal{S}_{N+1}(\delta')$ and $\mathcal{S}_N(\delta)$, whose derivation from Q_{N+1} and Q_N is far easier than the computation of $\tilde{Q}_{N+2}$ from Q_{N+1}.

As already mentioned, the iteration involved in the optimality problem allows to compute nondecreasing converging lower bounds to the optimal cost. The lower bound, together with the feasible upper bound, provides a measure of the accuracy of the approximation. In this context, deriving tighter upper bounds will reduce the number of iterations required for a desired accuracy and therefore the computational complexity of the whole approach.

9.5 Examples

In this section, we present two examples. In the first example, the sequence of sets Q_N^o converges after a finite number of steps. This allows us to compute the exact optimal cost. We also show that, for this example, an optimal strategy is a linear static full state feedback.

The second example has also been considered in [19]. It is known, [49], that an optimal full state linear controller for this problem must be dynamic. In this case, the sequence of Q_N^o does not seem to converge after few N. Thus, we derive upper and lower bounds to the optimal cost based on Theorem 9.4.2. A suboptimal controller can then be constructed using the method presented in Section 9.4.1.

9.5.1 Example 1

Consider the system:

$$x_{k+1} = \begin{pmatrix} 0 & 1 \\ 1 & 0 \end{pmatrix} x_k + \begin{pmatrix} 1 \\ 1 \end{pmatrix} w_k + \begin{pmatrix} 0 \\ 1 \end{pmatrix} u_k$$
$$z_k = \begin{pmatrix} 1 & 0 \\ 0 & 1 \end{pmatrix} x_k.$$

We solve the optimality problem for increasing N following the recursion given by Theorem 9.3.1. According to assumption $A3$, Q_0 is initialized to the set $\{x \mid \|C_1 x\|_\infty \le 1\}$. Thus, Q_0^o is initialized to be the unit diamond in R^2.

For $N \ge 2$, the polytopes Q_N^o stop changing, $Q_N^o = Q_{N+1}^o$. The vertices of Q_2^o are

$$vertQ_2^o = \left\{ \pm \begin{pmatrix} 1 \\ 0 \\ 0 \end{pmatrix}, \pm \begin{pmatrix} 0 \\ 1 \\ 1 \end{pmatrix}, \pm \begin{pmatrix} 0 \\ 1 \\ -1 \end{pmatrix}, \pm \begin{pmatrix} 0 \\ 0 \\ 2 \end{pmatrix} \right\}$$

Thus, the optimal performance is given by

$$\gamma^o = J_2(0,\delta) = \max_{q \in Q_2^o} q^T \begin{pmatrix} 0 \\ \delta \end{pmatrix} = \left\| \begin{pmatrix} 1 & 0 & 0 \\ 0 & 1 & 1 \\ 0 & 1 & -2 \\ 0 & 0 & 2 \end{pmatrix} \begin{pmatrix} 0 \\ 0 \\ \delta \end{pmatrix} \right\|_\infty = 2\delta.$$

Hence, given $\delta \ge 0$, there is an optimal full state strategy such that the worst-case amplitude of the output is twice the amplitude of the input. As already mentioned, in general, the optimal strategy may depend on δ. Following the development in Section 9.3.2, from Equation (9.2), and the fact that $Q_N^o = Q_{N+1}^o$ for $N \ge 2$, we have that

$$J_\infty(x_1, x_2, \delta) = \min_u \left\| \begin{bmatrix} 1 & 0 & 0 \\ 0 & 1 & 0 \\ 1 & 0 & 2 \\ 1 & 0 & 0 \\ 0 & 0 & 2 \\ 0 & 1 & -1 \\ 1 & 0 & 0 \\ 1 & 0 & -2 \\ 0 & 0 & 2 \end{bmatrix} \begin{bmatrix} x_1 \\ x_2 \\ \delta \end{bmatrix} + \begin{bmatrix} 0 \\ 0 \\ 1 \\ 1 \\ 0 \\ 0 \\ 1 \\ 1 \\ 0 \end{bmatrix} u \right\|_\infty$$

where $V_{S_2^o}\overline{A}$ and $V_{S_2^o}\overline{B}$ have been made explicit.

From Equation (9.3), the optimal u is given by

$$u^o = arg \min_u \left\| \begin{bmatrix} 1 & 0 & 2 \\ 1 & 0 & 0 \\ 1 & 0 & 0 \\ 1 & 0 & -2 \end{bmatrix} \begin{bmatrix} \chi \\ \delta \end{bmatrix} + \begin{bmatrix} 1 \\ 1 \\ 1 \\ 1 \end{bmatrix} u \right\|_\infty$$

It is easy to verify that the optimal u is given by $u^o = -x_1$, which is independent of δ. Thus, we have shown that there is an optimal infinite horizon stationary full state feedback strategy which is linear and static.

9.5.2 Example 2

Consider the following system

$$x_{k+1} = \begin{pmatrix} 0 & 1 & 0 \\ 0 & 0 & 1 \\ 4.6 & -23.5 & 2.7 \end{pmatrix} x_k + \begin{pmatrix} 0 \\ 0 \\ 1 \end{pmatrix} w_k + \begin{pmatrix} 0 \\ 0 \\ 1 \end{pmatrix} u_k$$

$$z_k = \begin{pmatrix} 1.51 & -2.5 & 1 \\ 0.5 & 0 & 0 \\ 0 & 0.5 & 0 \end{pmatrix} x_k$$

As already mentioned, this system has also been considered in [19], and it is based on a modification of the system presented in [49]. Its interesting property is that there is no full state linear static strategy which is optimal. In [19], it has been shown how to construct a nonlinear static controller that achieves performance arbitrarily closed to the performance of the optimal linear but dynamic controller. The optimal linear dynamic compensator for this problem ensures a performance $\gamma_{ld} = 3.0804$.

Here, we use the results of Section 9.4.1 to obtain, from the solution of a finite horizon optimality problem, upper and lower bounds to the optimal infinite horizon cost, and a suboptimal stabilizing control strategy.

Once again, we have solved the optimality problem for increasing N following the recursion given by Theorem 9.3.1, and, according to assumption $(A3)$, Q_0^o was initialized to the unit diamond in R^3. We stopped the recursion at $N = 9$. Q_9^o has 81 vertices, and $J_9(0, \delta) = 2.5309\delta$. Thus 2.5309 is a lower bound on the cost of the infinite horizon problem.

We should mention that B_1 does not have rank n_s. Although sufficient, this condition is not necessary to guarantee that the assumption of Lemma 9.4.5 is satisfied. In this example, we could verify that the assumption of Lemma 9.4.5 holds.

To compute the upper bound we have picked $\delta = 0.3246 = 1/3.0809$. This δ, selected after few trials, provides the smallest upper bound on the achievable performance we could find. For this δ, the inclusion in (9.9) is satisfied with $\delta' = .9999\delta$. Thus, an upper bound for the infinite horizon optimal cost is $\gamma = 3.0812$, very close to γ_{ld}. This level of performance is achieved by the control strategy constructed as described in Section 9.4.1.

Through simulation of the closed loop response to pseudo-random binary sequences of length 1500, uniformly distributed between ± 1 , we have verified that the maximum amplitude of the output vector did not exceed γ.

Summary and Comments. In this chapter, we have proposed a dynamic programming approach to the optimal ℓ_1 full state feedback . The main advantage of this approach is the direct derivation of a, generally, nonlinear optimal control law. We have considered both finite and infinite horizon cases, and studied the feasibility and the optimality problems. Once again, looking at the dual problem has provided us with important extra information, we have been able to characterize each problem the as polytopic functions recursively generated. From the analysis of the polytopes generated at each step, and under some assumptions, we have shown the existence of a stabilizing strategy for the infinite horizon problem, and presented two methods to derive suboptimal strategies when the optimal strategy is too complex to compute. Since the proposed approach is more general than the ones proposed in the literature, we think that eventually it could lead to more general and powerful results. In particular, research is undergoing to remove part of the assumptions used for the infinite horizon problem, especially assumptions $A1 - A3$, that guarantee both the BIBO stability of the closed loop and that the sets in the recursion are nested. We are also investigating the issue of deriving tighter upper bounds.

10. Conclusions

Exact solutions to practical control problems can often be computationally intractable. Consequently, computable approximate solutions must be found and information about the optimal solution should be recovered. In this monograph, we have shown that duality theory is an essential tool for deriving *informative* computational methods for convex optimization problems.

We have derived duality theory results and applied them to the study of several multi-objective control problems which are equivalent to convex constrained (generally infinite dimensional) optimization problems. We have shown that, by an appropriate selection of the positive cone, such problems are equivalent to abstract linear programming problems. This has allowed us to develop a uniform treatment for the problems of interest. The given convex optimization problem is first transformed into a linear programming problem. Then, duality theory is applied to the linear program to construct the dual program and study its properties. Finally, the dual convex problem is obtained, by transforming back the abstract dual linear program into the equivalent convex optimization problem. Often, different optimization problems can represent the same multi-objective control problem. The choice of the space where the problem lies, the topology on the space, and the positive cone can be fundamental in determining the properties of the duality relationship of the associated primal-dual pairs. Some properties are particularly important to the computation of exact or approximate solutions to the problem. A basic requirement is that the primal-dual pair has no duality gap. Another desired property of the pair is that finite support feasible solutions can approximate arbitrarily well the optimal solutions of both the primal and the dual problem. A stronger property is that these finite support approximating solutions can be computed via finite dimensional optimization problems. These properties allow the development of computational schemes that provide suboptimal feasible solutions with any desired accuracy. These properties of the primal-dual pair have been investigated in the study of several multi-objective problems. The analysis of the solvability of these problems has been rather complete. These results are the basis of a CAD methodology that allows direct performance specifications and provides the designer with structural information on the optimal design and its limitations.

Based on the solution of a special multi-objective problem we have derived a new computational method for ℓ_1, superior to the existing methods. By solving a sequence of finite dimensional LMI problems, we obtain converging upper and lower bounds to the optimal cost and suboptimal solutions converging in norm to the optimal ℓ_1 solution. Moreover, this method does not require the computation of the interpolation conditions and does not require the reordering of input and outputs as the Delay augmentation does.

The issues of deriving exact or approximate solutions are similar when the problems are posed as dynamic games in state space. We have studied the problem of finding the state feedback controller that minimizes the worst-case peak-to-peak amplification of the closed loop system. Once again, duality theory provides important extra information about the problem and allows the derivation of the structure of the optimal strategy and of approximation methods, when the optimal strategy cannot be computed exactly.

A. Proofs of Results in Chapter 5

Proof of Theorem 5.3.1

Notice that $FS \subset \ell_\infty$ and $FS \subset {\ell_\infty}^*$. Thus $\nu^o \leq \mu^o$, i.e., the optimal value of Problem (5.6) is less or equal than the optimal value of Problem (5.5) dual (with no duality gap) of Problem (5.4). Notice also that, if we restrict the domain of $\mathcal{A}_{temp}^*$ to FS instead of ${\ell_\infty}^*$, then $\mathcal{A}_{temp}^*|_{FS} = \mathcal{O}^* A_{temp}^T$.

Let FS_N be the subspace of FS consisting of all the finite support sequences zero for $k > N$: $FS_N = \{x \in FS \,|\, x(k) = 0, \;\; \text{for } k > N\}$.

In order to show that $\nu^o = \mu^o$, we first impose $z_2^* \in FS_N$ and show the connection with a finite-horizon problem with length of the horizon equal to N. Consider Problem (5.4) for any truncation of the horizon N. Let $P_N(\mathcal{A}_{temp})$ denote the resulting template operator.

$$\mu_N^o = \inf_{\Phi \in \ell_1^{n_z \times n_w}} \|\Phi\|_1 \tag{A.1}$$

subject to:

$$\mathcal{A}_{feas}\Phi = b_{feas}$$
$$P_N(\mathcal{A}_{temp})\Phi \leq P_N(b_{temp})$$

where $P_N : \ell_\infty \to R^N$ is a truncation operator. Given a sequence z, $P_N(z)$ returns a vector with the first N samples of z. From Theorem 5.2.2 we have that an optimal solution Φ_N^o to Problem (A.1) exists, and that

$$\mu_N^o = \sup \; \langle b_{feas}, w_1\rangle + \langle b_{temp}, w_2\rangle \tag{A.2}$$

subject to:

$$\|\mathcal{A}_{feas}^* w_1 + \mathcal{A}_{temp}^* P_N^*(w_2)\|_\infty \leq 1$$
$$w_2 \leq 0$$
$$w_1 \in c_0,\; w_2 \in R^T$$

From the remarks following Theorem 5.2.2, we also know that the optimal cost of the above problem does not change if we consider $w_1 \in FS$ instead of c_0. Since $P_N^*(w_2)$ belongs to FS_N, then, μ_N^o is also the optimal cost of the following problem:

$$\mu_N^o = \sup \; \langle b_{feas}, w_1\rangle + \langle b_{temp}, w_2\rangle \tag{A.3}$$

subject to:

$$\|\mathcal{A}_{feas}^* w_1 + \mathcal{A}_{temp}^* w_2\|_\infty \leq 1$$
$$w_2 \leq 0$$
$$w_1 \in FS,\; w_2 \in FS_N$$

which closely resembles Problem (5.6).

As the truncated horizon increases, the resulting primal problems are more and more constrained, thus the sequence of optimal costs $\{\mu_N^o\}$ is monotonically nondecreasing. It is convergent, since it is bounded above by μ^o. Denote by μ^* its limit point. Note that $\mu^* \leq \mu^o$, since for any finite N, $\mu_N^o \leq \mu^o$.

The sequence of optimal solutions, $\{\Phi_N^o\}$, is uniformly bounded in $\ell_1^{n_z \times n_w} = (c_0^{n_z \times n_w})^*$, thus, from Alaoglu's Theorem ([23]), there is a *weak** convergent subsequence $\{\Phi_{N_s}^o\}$. (If X is separable then a subset $S \subset X^*$ is *weak** compact if and only if it is *weak** sequentially compact ([55] pg. 346)). Let Φ^{w^*} denote its *weak** limit. We are now going to show that Φ^{w^*} is feasible for Problem (5.4) and therefore optimal. To simplify the notation, let

$$\mathcal{A}_N = \begin{bmatrix} \mathcal{A}_{feas} \\ P_N(\mathcal{A}_{temp}) \end{bmatrix}$$

$$b_N = \begin{bmatrix} b_{feas} \\ P_N(b_{temp}) \end{bmatrix}.$$

For any finite N, the set of equality and inequality constraints:

$$\begin{array}{rcl} \mathcal{A}_{feas}\Phi & = & b_{feas} \\ P_N(\mathcal{A}_{temp})\Phi & \leq & P_N(b_{temp}) \end{array}$$

is denoted, for brevity, with $\mathcal{A}_N\Phi \leq b_N$.

From the definition of *weak** convergence, we have that, for any $y \in c_0^{n_z \times n_w}$, $\langle y, \Phi_{N_s}^o \rangle \to \langle y, \Phi^{w^*} \rangle$ as N_s goes to ∞. Since each row of $\mathcal{A}_N$ is an element in c_0, then, for each $i \geq 1$, we have that $(\mathcal{A}_N\Phi)_i = (\mathcal{A}_N\mathcal{O}\Phi)_i = \langle \mathcal{O}^*(\mathcal{A}_N)_i^T, \Phi \rangle$ with $\mathcal{O}^*(\mathcal{A}_N)_i^T \in c_0^{n_z \times n_w}$. From the *weak** convergence it follows that, for any fixed $M \geq 1$,

$$\mathcal{A}_M\Phi_{N_s}^o \to \mathcal{A}_M\Phi^{w^*}$$

as N_s goes to infinity, where the convergence is intended pointwise. Given that, for all $N_s \geq M$, , $\Phi_{N_s}^o$ satisfies the constraints $\mathcal{A}_M\Phi_{N_s}^o \leq b_M$, then, also $\mathcal{A}_M\Phi^{w^*} \leq b_M$ must hold for any M. Thus, Φ^{w^*} is feasible, since

$$\limsup_{M\to\infty} \mathcal{A}_M\Phi^{w^*} \leq \limsup_{M\to\infty} b_M = b_\infty.$$

By looking at each $\Phi_{N_s}^o$ as a bounded linear operator from $c_0^{n_z \times n_w}$ to R, the *weak** convergence of $\{\Phi_{N_s}^o\}$ is equivalent to the strong operator convergence of $\Phi_{N_s}^o$, with limit Φ^{w^*}. Thus, from Lemma 4.9 – 5 pg. 267 [24], it follows that

$$\|\Phi^*\|_1 \leq \liminf_{s\to\infty} \|\Phi_{N_s}^o\|_1 \leq \mu^o.$$

Therefore, since Φ^* is feasible, all the inequalities above are in fact equalities and Φ^* is an optimal solution for (5.4). If $\Phi^o = \Phi^*$ is unique, then, all the subsequences must converge *weak** to it, so that the whole sequence converges *weak** to it. The rest of the results now follows from the convergence of μ_N^o to μ^o. ∎

Proof of Theorem 5.3.2

Any Φ satisfying $\mathcal{A}_{feas}\Phi = b_{feas}$ can be written by using the Youla parametrization as:

$$\Phi = H_p - UQ$$

for some $Q \in \ell_1^{n_u \times n_y}$ and U with finite support, and $H_p = \Phi_p$. To see this, let $\Phi = \bar{H} - \bar{U}\bar{Q}$ have a general form with $\bar{H}$ and $\bar{U}$ not of finite support. Since Φ_p is feasible, we have that $\Phi_p = \bar{H} - \bar{U}Q_p$, and we can rewrite the parametrization as $\Phi = \Phi_p - \bar{U}Q_1$. Finally, since Q_1 can cancel any stable pole of $\bar{U}$ and replace it with a pole at the origin, we have that for any stable Q_1, there is a stable Q such that $\bar{U}Q_1 = UQ$ with $\hat{U}(\lambda)$ polynomial, i.e., U of finite support.

Consider an optimal solution Φ^o for Problem (5.4). We know from theorem 5.3.1 that Φ^o exists. $\Phi^o = H_p - UQ^o$ for some Q^o. Notice however, that the existence of the optimal solution is not required in the proof. In cases where the optimal solution does not exist, for any $\epsilon > 0$, one can always select a feasible Φ with $\|\Phi\|_1 - \mu^o < \epsilon$, and use such Φ instead of Φ^o. Define the following problem:

$$\nu^o = \inf_{Q \in FS^{n_z \times n_w}} \|UQ - UQ^o\|_1 \tag{A.4}$$

subject to:

$$\mathcal{A}_{temp}UQ \geq \mathcal{A}_{temp}H_p - b_{temp}$$

It is easy to see that if $\nu^o = 0$ then Equation (5.8) holds. ν^o is the minimum distance between the optimal solution Φ^o and any finite support feasible solution. We have used the Youla parametrization instead of the feasibility constraints $\mathcal{A}_{feas}\Phi = b_{feas}$.

To compute ν^o we derive and solve the dual of Problem (A.4). The generalized linear program equivalent to Problem (A.4) is posed in the space $X = \ell_1^{n_u \times n_w} \times \ell_1^{n_z \times n_w} \times \ell_\infty \times R$ and is given by:

$$\nu^o \qquad \inf \qquad \gamma \tag{A.5}$$

subject to:

$$\begin{bmatrix} U & -I & 0 & 0 \\ \mathcal{A}_{temp} & 0 & I & 0 \end{bmatrix} \begin{bmatrix} Q \\ \Psi \\ \xi \\ \gamma \end{bmatrix} = \begin{bmatrix} UQ^o \\ \mathcal{A}_{temp}H_p - b_{temp} \end{bmatrix}$$

$$(Q, \Psi, \xi, \gamma) \in -P$$

where

$$-P = \{(Q, \Psi, \xi, \gamma) \in X \mid Q \in FS^{n_u \times n_w}, \|\Psi\|_1 - \gamma \leq 0, \xi \leq 0, \gamma \geq 0\}$$

We leave to the reader to verify that, given the assumptions and the fact that the operator of equality constraints has closed range in $\ell_1^{n_z \times n_w} \times \ell_\infty$, Theorem 4.2.2 applies. Since the conjugate positive cone is given by

$$P^\oplus = \{(Q^*, \Psi^*, \xi^*, \gamma^*) \in X^* \mid Q^* = 0,\ \|\Psi^*\|_\infty + \gamma^* \leq 0,\ \xi^* \geq 0,\ \gamma^* \leq 0\}.$$

Then from Theorem 4.2.2 and some rearrangements, the dual of Problem (A.4) is given by:

$$\begin{aligned} \nu^o = \quad & \max \quad \langle UQ^o, z_1^* \rangle + \langle A_{temp} H_p - b_{temp}, z_2^* \rangle \\ & \text{subject to:} \\ & -U^* z_1^* - U^* A_{temp}^* z_2^* = 0 \\ & \| - z_1 \|_\infty \leq 1 \\ & z_2^* \geq 0 \\ & z_1^* \in \ell_\infty,\ z_2^* \in {\ell_\infty}^* \end{aligned}$$

Substituting the constraints in the cost we have that

$$\langle UQ^o, z_1^* \rangle = \langle Q^o, U^* z_1^* \rangle = -\langle Q^o, U^* A_{temp}^* z_2^* \rangle = -\langle A_{temp} UQ^o, z_2^* \rangle$$

Thus the dual problem becomes the following:

$$\begin{aligned} \nu^o = \quad & \max \quad \langle A_{temp}(H_p - UQ^o) - b_{temp}, z_2^* \rangle \\ & \text{subject to:} \\ & -U^* z_1^* - U^* A_{temp}^* z_2^* = 0 \\ & \| - z_1 \|_\infty \leq 1 \\ & z_2^* \geq 0 \\ & z_1^* \in \ell_\infty,\ z_2^* \in {\ell_\infty}^* \end{aligned} \tag{A.6}$$

But, since the optimal solution Φ^o is feasible, $A_{temp}(H_p - UQ^o) - b_{temp} \leq 0$. Thus the cost in Problem (A.6) is always less then or equal to zero for any feasible (z_1^*, z_2^*). Since we know that there is no duality gap between Problem (A.4) and Problem (A.6), it follows that ν^0 must be equal to zero. ■

B. Proofs of Results in Chapter 6

Proof of Theorem 6.2.1

Notice that Problem (6.7) is similar to Problem (5.1). In both problems, the positive cone defining the inequality constraints has nonempty interior. The same argument as in Theorem 5.2.1 allows us to conclude that the dual, with no duality gap, of Problem (6.7) is given by:

$$\begin{aligned} &\max && \langle b_{feas}, z_1^* \rangle + \langle \gamma, z_2^* \rangle \\ &\text{subject to:} && \\ & && \|\mathcal{A}_{feas}^* z_1^* + \mathcal{A}_{\mathcal{H}_\infty}^* z_2^*\|_\infty \leq 1 \\ & && z_2^* \leq 0 \\ & && z_1^* \in \ell_\infty,\ z_2^* \in BV, \end{aligned} \tag{B.1}$$

where the cone $z_2^* \leq 0$ denotes the set of decreasing functions of bounded variations.

Recall that, from the definition of function of bounded variations, we have that $\langle x(\cdot), z_2^* \rangle = \int_0^{2\pi} x(\beta)\, d\, z_2^*(\beta)$.

Then, the dual of (6.7) can be readily written as follows:

$$\begin{aligned} &\max && \langle b_{feas}, z_1^* \rangle + \gamma \int_0^{2\pi} d\, z_2^*(\beta) \\ & \|\mathcal{A}_{feas}^* z_1^* + \mathcal{A}_{\mathcal{H}_\infty}^* z_2^*\|_\infty \leq 1 && \\ & z_2^* \leq 0 && \end{aligned} \tag{B.2}$$

First, notice that

$$\sqrt{C_1^2(z_2^*) + C_2^2(z_2^*)} \leq \left| \int_0^{2\pi} d\, z_2^*(\beta) \right| \text{ for any } z_2^* \in BV\ z_2^* \leq 0.$$

In fact, using the assumption that $z_2^* \leq 0$, we have that:

$$\begin{aligned} \sqrt{C_1^2(z_2^*) + C_2^2(z_2^*)} &= \left| \int_0^{2\pi} e^{-j\beta} d\, z_2^*(\beta) \right| \\ &\leq \int_0^{2\pi} \left| e^{-j\beta} \right| \cdot \left| d\, z_2^*(\beta) \right| \\ &= \|e^{-j\beta}\| \|z_2^*\| = \|z_2^*\| = -\int_0^{2\pi} d\, z_2^*(\beta) \end{aligned}$$

The equality holds when z_2^* is such that $d\,z_2^*(\beta) = \alpha\delta(\beta - \theta)$, where $\delta(\beta - \theta)$ is a Dirach's delta function, with $\alpha \in R$, in particular for $\alpha \leq 0$, and $\theta \in [0, 2\pi)$. Now define a new optimization problem:

$$\begin{array}{cc} \max & \langle b_{feas}, z_1^* \rangle - \gamma\sqrt{C_1^2(z_2^*) + C_2^2(z_2^*)} \\ \|\mathcal{A}_{feas}^* z_1^* + \mathcal{A}_{\mathcal{H}_\infty}^* z_2^*\|_\infty \leq 1 & \\ z_2^* \leq 0 & \end{array} \tag{B.3}$$

The optimal cost for this problem is the same as the cost of Problem B.2). To see this, notice that, since the feasibility sets for the two problems are the same and for any feasible z_2^* the cost functional of (B.3) is always greater than the cost functional of (B.2), then the optimal value of (B.2) is always smaller than the optimal value of (B.3). Moreover, let $z_{1,o}^*$ and $z_{2,o}^*$ achieve the maximum cost μ of (B.3) and let

$$\begin{array}{c} C_{1,o} = C_1(z_{2,o}^*) \\ C_{2,o} = C_2(z_{2,o}^*) \end{array}.$$

Then, from the previous argument it follows that, $z_1^* = z_{1,o}^*$ and z_2^* satisfying

$$d\,z_2^* = \alpha\delta(\beta - \theta), \tag{B.4}$$

with $\alpha = -\sqrt{C_{1,o}^2 + C_{2,o}^2}$ and $\theta = \tan^{-1}\dfrac{C_{2,o}}{C_{1,o}}$, are optimal for (B.2) with cost equal to μ.

Substituting (B.4) into Equations (6.5), it results that $C_1(z_2^*) = \alpha\cos(\theta)$ and $C_2(z_2^*) = \alpha\sin(\theta)$. Finally, the result follows if we let

$$C = \begin{bmatrix} \alpha\cos(\theta) \\ \alpha\sin(\theta) \end{bmatrix},$$

and we substitute this expression in (B.3). ∎

C. Proofs of Results in Chapter 8

Proof of Property (P3)

In here, we show that for any N Problem (8.2) is equivalent to a finite dimensional convex optimization in the first N elements of the impulse response of Q. This result follows from a property of the optimal $\mathcal{H}_2$ controller.

Consider the following problem

$$\begin{aligned} \mu_N^2 = \quad & \inf \quad \|\Phi_1\|_1^2 + \|\Phi_2\|_2^2 \qquad \text{(C.1)} \\ \text{subject to:} \quad & \\ & \Phi_1 = H_1 - U_1 Q_1 V \\ & \Phi_2 = H_2 - U_{12} Q_1 V - U_2 Q_2 V \\ & Q_1 \in R^{n_u \times n_y \times N}, Q_2 \in \ell_1^{n_u \times n_y} \end{aligned}$$

Notice that Q_1, Q_2 are minimizing solutions to Problem (C.1) if and only if they are minimizing solutions to Problem (8.2). Since Problem (C.1) is jointly convex in Q_1 and Q_2, it can be rewritten as follows:

$$\mu_N^2 = \inf_{\substack{\Phi_1 = H_1 - U_1 Q_1 V \\ Q_1 \in R^{n_u \times n_y \times N}}} \|\Phi_1\|_1^2 + \inf_{\substack{\Phi_2 = H_2 - U_{12} Q_1 V - U_2 Q_2 V \\ Q_2 \in \ell_1^{n_u \times n_y}}} \|\Phi_2\|_2^2 \qquad \text{(C.2)}$$

Given any $N \geq 0$, define

$$f^N(Q_1) = \inf_{\substack{\Phi_2 = H_2 - U_{12} Q_1 V - U_2 Q_2 V \\ Q_2 \in \ell_1^{n_u \times n_y}}} \|\Phi_2\|_2^2 \qquad \text{(C.3)}$$

Then the following is true.

Theorem C.0.1. *Consider the parametrization for Φ, $\hat{\Phi} = \hat{H} - \hat{U}\hat{Q}\hat{V}$ where $\hat{U}$ and $\hat{V}$ are inner (i.e., $\hat{U}^s\hat{U}$ and $\hat{V}\hat{V}^s$ are stable with stable inverse) and $\hat{H}$ is $\mathcal{H}_2$ optimal or equivalently, $\hat{U}^s\hat{H}\hat{V}^s \in \mathcal{RH}_2^\perp$. A parametrization with these properties can be obtained from the model-based optimal $\mathcal{H}_2$ controller. Then, for any $N \geq 0$,*

$$f^N(Q_1) = \|H_2 - U_{12} Q_1 V\|_2^2$$

i.e., the minimum is achieved at $Q_2 = 0$.

The result of this theorem implies that Problem (C.1) is equivalent to Problem (8.4)

Proof. Consider $\hat{\Phi}_2 = \hat{H}_2 - (U_1\hat{\underline{2}}Q_1)\hat{V} - \hat{U}_2\hat{Q}_2\hat{V}$, where $(U_1\hat{\underline{2}}Q_1)$ is the λ-transform of the seqeunce $U_{12}Q_1$. $\hat{\Phi}_2$ can be rewritten as follows

$$\hat{\Phi}_2 = \lambda^{-N}[\hat{H} - \hat{U}\hat{Q}\hat{V} - \hat{\Phi}_1$$

where $\Phi_1 = H_1 - U_1Q_1V$. Notice that $\hat{\Phi}_1(\lambda)$ is a polynomial in λ of order $N-1$. Assume without loss of generality that $\hat{U}^s\hat{U} = I$ and $\hat{V}\hat{V}^s = I$. Then we have that

$$\hat{U}^s\hat{\Phi}_2\hat{V}^s = \lambda^{-N}\hat{U}^s\hat{H}\hat{V}^s - \lambda^{-N}\hat{Q}_1 - \hat{Q}_2 - \lambda^{-N}\hat{U}^s\hat{\Phi}_1V^s$$

All the terms in the right hand side of the above equation are in $\mathcal{RH}_2^{\perp}$ with the exception of Q_2. Thus we have that

$$\begin{aligned}\|H_2 - U_{12}Q_1 - UQ_2V\|_2 &= \|\hat{U}^s(\hat{H}_2 - (U_1\hat{\underline{2}}Q_1)\hat{V} - \hat{U}\hat{Q}_2\hat{V})\hat{V}^s\|_2 \\ &= \|\lambda^{-N}\hat{U}^s\hat{H}\hat{V}^s - \lambda^{-N}\hat{Q}_1 - \lambda^{-N}\hat{U}^s\hat{\Phi}_1V^s\|_2 + \|\hat{Q}_2\|_2\end{aligned}$$

Therefore, the minimum value of $f^N(Q_1)$ is achieved by $Q_2 = 0$. Thus,

$$\min_{Q_2} \|H_2 - U_{12}Q_1 - UQ_2V\|_2 = \|H_2 - U_{12}Q_1V\|_2$$

■

Proof of Theorem 8.4.1

Consider the optimal solution, x^o, of the ℓ_1 problem, which exists under standard assumptions. The sequence $\{\|x^o\|_N\}$, converges to μ^o, and for each N, $\|x^o\|_N$ is an upper bound on μ_N. Thus, all the limit points of $\{\mu_N\}$ will have values greater than zero and less than μ^o.

Consider any subsequence $\{\mu_{N_n}\}$, converging to one of the limit points. To simplify the notation, we remove the subindex and denote the subsequence by $\{\mu_N\}$. Let μ^* be its limit. We are going to show that $\mu^* = \mu^o$.

Fix any $\epsilon > 0$. Since $\{\|x^o\|_N\}$ converges to μ^o, there exists an integer M_1 such that, for all $N \geq M_1$, $|\mu^o - \|x^o\|_N| < \epsilon$. Consider Problem (8.9). It is well known that there are finite support feasible dual sequences z^* whose cost, μ, is arbitrarily close to μ^o, say $\mu^o - \mu < \epsilon_1$ for $\epsilon_1 < \epsilon/2$. Since z^* is feasible, it is also true that $\|\mathcal{A}_{feas}^* z^*\|_\infty \leq 1$. Given $x^* = \mathcal{A}_{feas}^* z^*$, for any N, we have that

$$U_1^* P_N x^* V^* + U_{12}^* T_N x^* V^* = 0.$$

Now pick any $0 < \epsilon_2 < \min\{\epsilon/2, \mu\}$, and let $y^* = (1 - \epsilon_2/\mu)x^*$. Then $y^* \in \ell_2^{n_z \times n_w}$ and $\|y^*\|_\infty \leq 1 - \epsilon_2/\mu$. Moreover there exists an M_2 large enough such that

$$\|P_N y^*\|_\infty^2 + \|T_N y^*\|_2^2 \leq 1$$

for all $N \geq M_2$. Hence, for all $N \geq M_2$, y^* is feasible for Problem (8.6) with cost $\mu - \epsilon_2 < \mu_{M_2}$. Thus we have that, for all $N \geq \max\{M_1, M_2\}$,

$$\mu^o - \mu_N \geq \mu^o - \|x^o\|_N > -\epsilon$$

and

$$\mu^o - \mu_N < \mu^o - \mu + \epsilon_2 < \epsilon_1 + \epsilon_2 < \epsilon.$$

Therefore

$$|\mu^o - \mu_N| < \epsilon \qquad \text{for all } N \geq \max\{M_1, M_2\}.$$

Thus, for any $\epsilon > 0$ there exists an M such that $|\mu^o - \mu_N| < \epsilon$ for all $N \geq M$; hence, μ_N converges to μ^o. Since any convergent subsequence converges to μ^o, the whole sequence must converge to μ^o, and the result is proved.

■

Proof of Theorem 8.4.2

For each N, let $\Phi^N = [\Phi_1^N, \Phi_2^N]$ be the optimal primal solution, and $x^* = [x_1^*, x_2^*]$ the optimal dual solution. We know that they are aligned. Let $\mu_{1N} = \|\Phi_1^N\|_1$, $\mu_{2N} = \|\Phi_2^N\|_2$, $\gamma_1^N = \|x_1^*\|_\infty$, and $\gamma_2^N = \|x_2^*\|_2$. The alignment condition implies that:

$$\begin{aligned} \gamma_1^N &= \frac{\mu_{1N}}{\mu_N} \\ \gamma_2^N &= \frac{\mu_{2N}}{\mu_N}. \end{aligned} \tag{C.4}$$

Moreover $\Phi_2^N = \mu_N x_2^*$. Thus, $\|\Phi_1^N\|_2$ goes to zero if and only if γ_2^N goes to zero.

To derive a contradiction, assume that γ_2^N is not converging to 0. This implies that, there exist a positive constant $1 \geq a > 0$ such that for any positive integer M it is possible to find some $N_M \geq M$ for which

$$\gamma_2^{N_M} \geq a.$$

For simplicity, relabel the sequence $\gamma_2^{N_M}$ as γ_2^N. Then, from Lemma 8.4.1, also the sequence $\{\|\Phi_2^N\|_1\}$ is uniformly bounded by some constant $\alpha < \infty$. Since $\gamma_2^N \geq a$ for all N, we have that $\gamma_1^N \leq \sqrt{1-a^2} < 1$ for all N. Thus, from the alignment condition in (C.4), we have that $\|\Phi_1^N\|_1 \leq \mu_N\sqrt{1-a^2}$. The convergence of μ_N implies that the sequence $\{\|\Phi_1^N\|_1\}$ is uniformly bounded by some positive constant b. Thus, the sequence $\{\|\Phi^N\|_1\}$ is uniformly bounded because $\|\Phi^N\|_1 \leq \|\Phi_1^N\|_1 + \|\Phi_2^N\|_1$.

It follows from the Banach Alouglu theorem that there is a subsequence, $\{\Phi^{N_s}\}$, converging *weak** to some element Φ^{w^*}.

We claim that $\|\Phi^{w^*}\|_1 \leq \mu^o\sqrt{1-a^2}$. For each N, $\Phi_1^N \in R^{n_z \times n_w \times N}$. Φ_1^N can be seen as an element of $\ell_1^{n_z \times n_w}$ by considering it as an FIR sequence in $\ell_1^{n_z \times n_w}$. We still denote this extension as Φ_1^N. For any $\epsilon > 0$, there exists an N_0 such that $\|\Phi_1^{N_s}\|_1 \leq (\mu^o + \epsilon)\sqrt{1-a^2}$ for all $N_s \geq N_0$. Thus, for $N_s \geq N_0$ the sequence $\{\Phi_1^{N_s}\}$ contains a *weak** convergent subsequence $\{\Phi_1^{N_{sr}}\}$. Let $\Phi_1^{w^*}$ denote its *weak** limit point. Then $\Phi_1^{w^*} = \Phi^{w^*}$, since the sequence $\{\Phi_1^{N_{sr}} - \Phi^{N_{sr}}\}$ is *weak** convergent to the zero element. Moreover, we have that

$$\|\Phi^{w^*}\|_1 \leq (\mu^o + \epsilon)\sqrt{1-a^2}$$

Since ϵ can be arbitrarily small, we have that $\|\Phi^{w^*}\|_1 \leq \mu^o\sqrt{1-a^2}$.

We now show that Φ^{w^*} is a feasible solution to the ℓ_1 problem. This immediately implies that μ^o is not the optimal cost, and this contradiction will prove the assertion of the theorem.

Let $R : \ell_1^{n_u \times n_y} \rightarrow \ell_1^{n_z \times n_w}$ be the linear operator mapping Q to $RQ = UQV$. Under the current assumptions, R is one to one with closed range in $\ell_1^{n_z \times n_w}$. The proof of the above statement is left to the reader.

Let $Y^N = RQ^N = H - \Phi^N$. Notice that the ℓ_1 norm of Y^N is uniformly bounded since $\|Y^N\|_1 \leq \|H\|_1 + \|\Phi^N\|_1$ and $\|\Phi^N\|_1$ is converging to some $\mu \leq \mu^o\sqrt{1-a^2}$.

From Lemma 1 *pp*.155 in [22], we have that if R has closed range then there is a positive constant k such that, for any Y in the range of R, there is a Q satisfying $Y = RQ$, with $\|Q\|_1 \leq k\|Y\|_1$. Thus, if we consider the sequence, $\{Q^N\}$, of optimal Q's, we have also that the sequence of norms, $\{\|Q^N\|_1\}$ is uniformly bounded. Therefore, it contains a subsequence which is *weak** convergent to some $Q^{w^*} \in \ell_1^{n_u \times n_y}$, and moreover, $\Phi^{w^*} = H - UQ^{w^*}V$ since the sequence $\{\Phi^N - (H - UQ^N V)\}$ is *weak** convergent to zero. Summarizing, we have a feasible solution Φ^{w^*} with $\|\Phi^{w^*}\|_1 \leq \mu^o\sqrt{1-a^2} < \mu^o$. But this is impossible; hence, $\|\Phi_2^N\|_2$ must go to zeros for $N \rightarrow \infty$. ∎

D. Proofs of Results in Chapter 9

Proof of Theorem 9.3.1

We first show that the result is true for $k = 1$. For notational convenience, we drop the subscript $N-1$. $J_1(\chi, \delta)$ can be rewritten as follows:

$$J_1(\chi,\delta) = \min_u \max_{\|w\|_\infty \le \delta} \max \left\{ \begin{array}{c} \max_{r \in R_0^o} r^T (C_1\chi + D_{12}u) \\ \max_{q \in Q_0^o} q^T (A\chi + B_1 w + B_2 u) \end{array} \right\},$$

where we have used the facts that

$$J_0(\chi) = \max_{q \in Q_0^o} q^T \chi,$$

and

$$\|C_1\chi + D_{12}u\|_\infty = \max_{r \in R_0^o} r^T (C_1\chi + D_{12}u),$$

and R_0^o is the unit diamond in R^{n_z}.

Consider the following optimization:

$$f(\chi,\delta) = \max_{\|w\|_\infty \le \delta} \max \left\{ \begin{array}{c} \max_{r \in R_0^o} r^T (C_1\chi + D_{12}u) \\ \max_{q \in Q_0^o} q^T (A\chi + B_1 w + B_2 u) \end{array} \right\}. \tag{D.1}$$

We leave to the reader to verify that $f(\chi, \delta)$ can be equivalently expressed as follows:

$$f(\chi,\delta) = max \left\{ \begin{array}{c} \max_{r \in R_0^o} r^T (C_1\chi + D_{12}u) \\ \max_{q_1 \in Q_0^o} q_1^T (A\chi + B_2 u + B_1\delta) \\ \max_{q_2 \in Q_0^o} q_2^T (A\chi + B_2 u - B_1\delta) \end{array} \right\} \tag{D.2}$$

$f(\chi, \delta)$ is also given by the following optimization:

$$f(\chi,\delta) = \max_{p \in B^o} p^T \left[\begin{bmatrix} C_1 & 0 \\ V_{Q_0^o}A & V_{Q_0^o}B_1 \\ V_{Q_0^o}A & -V_{Q_0^o}B_1 \end{bmatrix} \begin{bmatrix} \chi \\ \delta \end{bmatrix} + \begin{bmatrix} D_{12} \\ V_{Q_0^o}B_2 \\ V_{Q_0^o}B_2 \end{bmatrix} u \right]$$

Using the definition of S_1^o we have that

$$f(\chi,\delta) = \max_{p\in S_1^o} p^T \left[\begin{bmatrix} C_1 & 0 \\ A & B_1 \\ A & -B_1 \end{bmatrix} \begin{bmatrix} \chi \\ \delta \end{bmatrix} + \begin{bmatrix} D_{12} \\ B_2 \\ B_2 \end{bmatrix} u \right]$$

Hence, we have that

$$J_1(x,\delta) = \inf_u f(\chi,\delta) = \inf_u \max_{p\in S_1^o} p^T \left[\begin{bmatrix} C_1 & 0 \\ A & B_1 \\ A & -B_1 \end{bmatrix} \begin{bmatrix} \chi \\ \delta \end{bmatrix} + \begin{bmatrix} D_{12} \\ B_2 \\ B_2 \end{bmatrix} u \right]$$

The above is a minimum norm problem, from standard duality theory results [22], we can write the above problem in the dual form and obtain that

$$J_1(\chi,\delta) = \max_{\substack{p\in S_1^o \\ p^T \begin{bmatrix} D_{12} \\ V_{Q_0^o} B_2 \\ V_{Q_0^o} B_2 \end{bmatrix} = 0}} p^T \begin{bmatrix} C_1 & 0 \\ V_{Q_0^o} A & V_{Q_0^o} B_1 \\ V_{Q_0^o} A & -V_{Q_0^o} B_1 \end{bmatrix} \begin{bmatrix} \chi \\ \delta \end{bmatrix}.$$

From the definition of $\tilde{A}$, $\tilde{B}$, and Q_1^o, we have that

$$J_1(\chi,\delta) = \max_{q\in Q_1^o} q^T \begin{bmatrix} \chi \\ \delta \end{bmatrix} \tag{D.3}$$

which is the desired result.

Now, consider

$$J_2(\chi,\delta) = \min_u \max_{\|w\|_\infty \le \delta} \max \left\{ \begin{matrix} \|C_1\chi + D_{12}u\|_\infty \\ J_1(A\chi + B_1 w + B_2 u, \delta) \end{matrix} \right\}$$

Substituting the expression (D.3), for J_1, we have that

$$J_2(\chi,\delta) = \min_u \max_{\|w\|_\infty \le \delta} \max \left\{ \begin{matrix} \max_{r\in R_0^o} r^T (C_1\chi + D_{12}u) \\ \max_{q\in Q_1^o} q^T \begin{bmatrix} A\chi + B_1 w + B_2 u \\ I\delta \end{bmatrix} \end{matrix} \right\}$$

Then, we can rewrite J_2 as follows:

$$J_2(\chi,\delta) = \min_u \max_{\|w\|_\infty \le \delta} \max \left\{ \begin{matrix} \max_{r\in R_0^o} r^T (C_1\chi + D_{12}u) \\ \max_{q\in Q_1^o} q^T \left(\begin{bmatrix} A \\ 0 \end{bmatrix} \chi + \begin{bmatrix} B_1 \\ 0 \end{bmatrix} w + \begin{bmatrix} B_2 \\ 0 \end{bmatrix} u + \begin{bmatrix} 0 \\ I \end{bmatrix} \delta \right) \end{matrix} \right\}$$

As for J_1, we can remove the dependence on w of J_2 and express J_2 as follows:

$$J_2(\chi,\delta) = \min_u \max \left\{ \begin{array}{c} \max\limits_{r\in R_0^o} r^T(C_1\chi + D_{12}u) \\ \max\limits_{q\in Q_1^o} q^T \left[\begin{bmatrix} A \\ 0 \end{bmatrix} \chi + \begin{bmatrix} B_1 \\ I \end{bmatrix} \delta + \begin{bmatrix} B_2 \\ 0 \end{bmatrix} u \right] \\ \max\limits_{s\in Q_1^o} s^T \left[\begin{bmatrix} A \\ 0 \end{bmatrix} \chi + \begin{bmatrix} -B_1 \\ I \end{bmatrix} \delta + \begin{bmatrix} B_2 \\ 0 \end{bmatrix} u \right] \end{array} \right\}$$

or, equivalently, as

$$J_2(\chi,\delta) = \min_u \max_{p\in\mathcal{B}^o} p^T \left[\begin{bmatrix} V_{R_0^o}C_1 & 0 \\ V_{Q_1^o}\begin{bmatrix} A \\ 0 \end{bmatrix} & V_{Q_1^o}\begin{bmatrix} B_1 \\ I \end{bmatrix} \\ V_{Q_1^o}\begin{bmatrix} A \\ 0 \end{bmatrix} & V_{Q_1^o}\begin{bmatrix} -B_1 \\ I \end{bmatrix} \end{bmatrix} \begin{bmatrix} \chi \\ \delta \end{bmatrix} + \begin{bmatrix} V_{R_0^o}D_{12} \\ V_{Q_1^o}\begin{bmatrix} B_2 \\ 0 \end{bmatrix} \\ V_{Q_1^o}\begin{bmatrix} B_2 \\ 0 \end{bmatrix} \end{bmatrix} u \right]$$

Using the definition of $\mathcal{T}$, J_2 can be rewritten as follows:

$$J_2(\chi,\delta) = \min_u \max_{p\in\mathcal{B}^o} p^T \mathcal{T}(V_{R_0^o}, V_{Q_1^o}) \left[\begin{bmatrix} C_1 & 0 \\ A & B_1 \\ A & -B_1 \\ 0 & I \end{bmatrix} \begin{bmatrix} \chi \\ \delta \end{bmatrix} + \begin{bmatrix} D_{12} \\ B_2 \\ B_2 \\ 0 \end{bmatrix} u \right]$$

From the properties of polar cones, we have that

$$J_2(\chi,\delta) = \min_u \max_{s\in S_2^o} s^T \left[\begin{bmatrix} C_1 & 0 \\ A & B_1 \\ A & -B_1 \\ 0 & I \end{bmatrix} \begin{bmatrix} \chi \\ \delta \end{bmatrix} + \begin{bmatrix} D_{12} \\ B_2 \\ B_2 \\ 0 \end{bmatrix} u \right]$$

where $S_2^o = \mathcal{T}(V_{R_0^o}, V_{Q_1^o})^T \mathcal{B}^o$. Finally, the above is a minimum norm problem whose dual is

$$\begin{aligned} J_2(\chi,\delta) = \quad & \max_{s\in S_2^o} \quad s^T \begin{bmatrix} C_1 & 0 \\ A & B_1 \\ A & -B_1 \\ 0 & I \end{bmatrix} \begin{bmatrix} \chi \\ \delta \end{bmatrix} \\ & s^T \begin{bmatrix} D_{12} \\ B_2 \\ B_2 \\ 0 \end{bmatrix} = 0 \end{aligned}$$

From the definition of Q_2^o we obtain that,

$$J_2(\chi,\delta) = \max_{q\in Q_2^o} q^T \begin{bmatrix} \chi \\ \delta \end{bmatrix}.$$

The result of the theorem now follows by applying the same argument recursively. ■

Proof of Theorem 9.3.2

First, notice that the result of the theorem trivially holds for $N = 0$. Now consider H_1. Dropping the subscript $N-1$ and using the expression for J_0, H_1 can be rewritten as follows:

$$H_1(\chi) = \min_u \max \left\{ \begin{array}{c} \|C_1\chi + D_{12}u\|_\infty \\ \\ \max\limits_{\|w\|_\infty \le \delta'} \max\limits_{q \in Q_0^o} q^T(A\chi + B_2u + B_1w) \end{array} \right\}$$

Thus, $H_1(\chi) < 1$ if and only if there exists some u such that,

$$\begin{array}{c} \|C_1\chi + D_{12}u\|_\infty < 1 \\ \text{and} \\ \max\limits_{\|w\|_\infty \le \delta'} \max\limits_{q \in Q_0^o} q^T(A\chi + B_2u + B_1w) < 1 \end{array}$$

or, equivalently, $H_1(\chi) < 1$ if and only if

$$\begin{array}{c} \max\limits_{r \in R_0^o} r^T(C_1\chi + D_{12}u) < 1 \\ \text{and} \\ \max\limits_{\|w\|_\infty \le \delta'} \max\limits_{q \in Q_0^o} q^T(A\chi + B_2u + B_1w) < 1 \end{array} . \tag{D.4}$$

Next we remove the direct dependence on w form the above conditions. Consider the second condition,

$$\max_{\|w\|_\infty \le \delta'} \max_{q \in Q_0^o} q^T(A\chi + B_2u + B_1w) < 1.$$

$q^T(A\chi + B_2u + B_1w)$ is a convex function of q. Thus, its maximum is achieved by some of the vertices q_i's of Q_0^o. We have the following implications:

$$\begin{array}{ll} & \max\limits_{\|w\|_\infty \le \delta'} \max\limits_{q \in Q_0^o} q^T(A\chi + B_2u + B_1w) < 1 \\ \Leftrightarrow & q_i^T(A\chi + B_2u + B_1w) < 1, \quad \left\{ \begin{array}{l} \forall q_i \in vertQ_0^o \\ \forall w : \|w\|_\infty \le \delta' \end{array} \right. \\ \Leftrightarrow & q_i^T(A\chi + B_2u) + \|q_i^T B_1\|_1 \delta' < 1 \quad \forall q_i \in vertQ_0^o \\ \Leftrightarrow & \left\{ \begin{array}{c} \dfrac{q_i^T}{1 - \|q_i^T B_1\|_1 \delta'}(A\chi + B_2u) < 1 \\ \text{and} \\ \|q_i^T B_1\|_1 \delta' < 1 \end{array} \right\}, \quad \forall q_i \in vertQ_0^o \\ \Leftrightarrow & \left\{ \begin{array}{c} \|q_i^T B_1\|_1 \delta' < 1, \quad \forall q_i \in vertQ_0^o \\ \text{and} \\ \max\limits_{s \in S_0^o} s^T(A\chi + B_2u) < 1 \end{array} \right. \end{array}$$

$$\text{where } S_0^o = conv\left\{\frac{q_i^T}{1-\|q_i^T B_1\|_1\delta'}, \mid q_i \in vertQ_0^o, \text{ and } \|q_i^T B_1\|_1\delta' < 1\right\}.$$

From the previous argument, it follows that $H_1(\chi) < 1$ if and only if

$$\|q_i^T B_1\|_1\delta' < 1 \ \forall q_i \in vertQ_0^o \tag{D.5}$$

and

$$\min_u \left\{\begin{array}{l} \max\limits_{r\in R_0^o} r^T(C_1\chi + D_{12}u) \\ \max\limits_{s\in S_0^o} \ s^T(A\chi + B_2u) \end{array}\right\} < 1. \tag{D.6}$$

We now concentrate our attention on the optimization problem in (D.6). Notice that this problem does not depend on w, since now the dependence on w is included into the Condition (D.5).

From the definition of P_k^o, $\tilde{A}$, and $\tilde{B}$ in the theorem statement, the optimization in (D.6) can be rewritten as follows:

$$\min_u \max_{p\in P_0^o} p^T\left[\tilde{A}\chi + \tilde{B}u\right]$$

Similarly to the proof of Theorem 9.3.1, the above optimization can be shown to be equivalent to the following problem

$$\begin{array}{ll} \max & p^T\tilde{A}\chi < 1 \\ p\in P_0^o & \\ p^T\tilde{B} = 0 & \end{array} \tag{D.7}$$

Finally, using a property of polar cones, Problem (D.7) can be rewritten as

$$\max_{q\in Q_1^o} q^T\chi$$

where

$$Q_1^o = \tilde{A}^T\left\{p \in P_0^o \mid p^T\tilde{B} = 0\right\}.$$

Thus we obtain that $H_1(\chi) < 1$ if and only if

$$\begin{array}{c} \|q_i^T B_1\|_1\delta' < 1 \ \forall q_i \in vertQ_0^o \\ \text{and} \\ H_1(\chi) = \max\limits_{q\in Q_1^o} q^T\chi < 1 \end{array}$$

We can now apply the same argument recursively, and obtain the desired result. ■

Proof of Lemma 9.4.1

First, notice that the set Q_1 can be defined as follows:

$$Q_1 = \{(x,\delta) \mid \exists u \text{ such that } \|C_1 x + D_{12}u\|_\infty \le 1, \\ \text{and } Ax + B_2 u + B_1 w \in Q_0 \, \forall w \text{ with } \|w\|_\infty \le \delta \}$$

Now, if $(x,\delta) \in Q_1$ then, from the definition of Q_0, $x \in Q_0$. Denote by $Q_1(\delta)$ the set of x such that $(x,\delta) \in Q_1$. Note that $Q_1(\delta)$ can be empty. Then, it follows that $Q_1(\delta) \subset Q_0$ for any $\delta \ge 0$.

For $k > 1$, Q_{k+1} can be defined as follows

$$Q_{k+1} = \{(x,\delta) \mid \exists u \text{ such that } \|C_1 x + D_{12}u\|_\infty \le 1, \\ \text{and } (Ax + B_2 u + B_1 w \,,\, \delta) \in Q_k \, \forall w \text{ with } \|w\|_\infty \le \delta \}.$$

In particular for $k = 2$, if $(x,\delta) \in Q_2$, then δ belongs to Q_1 and there exists a u such that

$$\begin{array}{l} \|C_1 x + D_{12}u\|_\infty \le 1 \\ Ax + B_2 u + B_1 w \in Q_1(\delta) \subset Q_0 \quad \forall w : \|w\|_\infty \le \delta \end{array}$$

Thus, if $(x,\delta) \in Q_2$, there exists a u such that

$$\begin{array}{l} \|C_1 x + D_{12}u\|_\infty \le 1 \\ Ax + B_2 u + B_1 w \in Q_0 \quad \forall w : \|w\|_\infty \le \delta \end{array}$$

or equivalently, $(x,\delta) \in Q_1$. Repeating the same argument for $k > 2$ shows that $Q_{k+1} \subset Q_k$.

Regarding the compactness, notice that the sets Q_k's are clearly closed. Since they are nested, to prove that they are compact, it is enough to show that Q_1 is bounded in R^{n_s+1} for any $k \ge 1$.

We first show that Q_0 is bounded. One way to see this is to show that

$$J_0(x) = \min_u \|C_1 x + D_{12}u\|_\infty$$

is a norm on $x \in R^{n_s}$. Since C_1 has rank n_s and $D_{12}^T C_1 = 0$, we have that $J_0(x) = 0 \Leftrightarrow x = 0$.

The triangle inequality can be proved by using the dual representation. Let R_0 be the unit ℓ_∞-ball, then

$$\begin{aligned} J_0(x_1 + x_2) &= \min\nolimits_u \|C_1(x_1+x_2) + D_{12}u\|_\infty \\ &= \max_{\substack{r \in R_0^o \\ r^T D_{12} = 0}} r^T C_1 (x_1 + x_2) \\ &\le \max_{\substack{r \in R_0^o \\ r^T D_{12} = 0}} r^T C_1 x_1 + \max_{\substack{r \in R_0^o \\ r^T D_{12} = 0}} r^T C_1 x_2 \\ &= J_0(x_1) + J_0(x_2) \end{aligned}$$

Finally, it is immediate that $J_0(\alpha x) = |\alpha| J_0(x)$ for any $\alpha \in R$. Thus, Q_0 is bounded since $Q_0 = \{x \mid g(x) \le 1\}$.

Consider now Q_1. Previously we have shown that for any $\delta \ge 0$,

$$Q_1(\delta) \subset Q_0.$$

Thus, for any $\delta \geq 0$, $Q_1(\delta)$ is bounded. To show that Q_1 is bounded, we only have to show that the set

$$\{\delta \mid (x, \delta) \in Q_1, \text{ for some } x\}$$

is bounded. But, this follows from the fact that Q_0 is bounded and the assumption on the rank of B_1. The details are left to the reader. Thus Q_1 is compact and, hence, any Q_k for $k \geq 1$ is compact being a closed subset of Q_1. ∎

References

1. M. A. Dahleh, I. J. Diaz-Bobillo. *Control of Uncertain Systems: A Linear Programming Approach*. Prentice-Hall, New Jersey, 1995.
2. B. A. Francis. *A Course in $\mathcal{H}_\infty$ Control Theory*. Springer-Verlag, 1987 (Lecture Notes in Control and Information Science volume no.88)
3. J. M. Maciejowski. *Multivariable Feedback Design*. Addison-Wesley, London, 1989.
4. A. Megretski, A. Rantzer. *System Analysis via Integral Quadratic Constraints*. IEEE Transactions on Automatic Control, 1997; 42:819
5. F. Paganini. *State Space Conditions for Robust $\mathcal{H}_2$ Analysis*. Proceedings American Control Conference, 1997, Vol. 2 1230
6. K. Y. Yang, S. R. Hall, E. Feron. *A new design method for robust $\mathcal{H}_2$ controllers using Popov multipliers*. Proceedings American Control Conference, 1997, pp. 1235-1240
7. S. P. Boyd, C. H. Barratt. *Linear controller design : limits of performance*. Prentice Hall, New Jersey, 1991.
8. S. P. Boyd, L. El Ghaoui, E. Feron, V. Balakrishnan. *Linear Matrix Inequalities in System and Control Theory*. Philadelphia: Society for Industrial and Applied Mathematics, 1994, SIAM Studies in Applied Mathematics 15.
9. G. Deodhare. *Design of Multivariable Linear Systems Using Infinite Linear Programming*. Ph.D. Thesis, University of Waterloo, Ontario, Canada, 1990.
10. P. Voulgaris. *Optimal $\mathcal{H}_2/\ell_1$ control: the SISO case*. Proceedings of the 33rd Conference on Decision and Control, 1994, pp. 3181-3186.
11. H. Rotstein, A. Sideris. *$\mathcal{H}_\infty$ Optimization with Time Domain Constraints*. IEEE Transactions on Automatic Control, 1994; 39:762-770
12. M. Sznaier. *Mixed $\ell_1/\mathcal{H}_\infty$ Controllers for MIMO Discrete-Time Systems*. Proceedings 33rd Conference on Decision and Control, 1994, pp. 3187-3191
13. P.P. Khargonekar, M. A. Rotea. *Mixed $\mathcal{H}_2/\mathcal{H}_\infty$ Control: A Convex Optimization Approach*. IEEE Transactions on Automatic Control, 1991; 36:824
14. J. S. McDonald, J. B. Pearson. *ℓ_1-Optimal Control of Multivariable Systems with Output Norm Constraints*. Automatica, 1991; 27:317-329
15. O. J. Staffans. *On the Four-block Model Matching Problem in ℓ_1 and Infinite-Dimensional Linear Programming*. SIAM J. Control and Optimization, 1993; 31;747–779
16. I. J. Diaz-Bobillo, M.A. Dahleh. *Minimization of the Maximum Peak-to-Peak Gain: The General Multiblock Problem*. IEEE Transactions on Automatic Control, 1993; 38:1459-1483
17. A. A. Stoorvogel. *Nonlinear L_1 Optimal Controllers for Linear Systems*. IEEE Transactions on Automatic Control, 1995; 40:694-696
18. J. S. Shamma. *Nonlinear state feedback for ℓ_1 optimal control*. Systems and Control Letters, 1993; 21:265-270.
19. J. S. Shamma. *Optimization of the ℓ_∞ induced norm under full state feedback*. IEEE Transactions on Automatic Control, 1996; 41:533-544

20. N. Elia, P. M. Young, M. A. Dahleh. *Multiobjective Control via Infinite-Dimensional LMI Optimization.* Proceedings 33rd Annual Allerton Conference on Communication, Control, and Computing, 1995; 186-195
21. P. M. Young, M. A. Dahleh. *Infinite Dimensional Convex Optimization in Optimal and Robust Control.* IEEE Transactions on Automatic Control, 1997; 42:1370.
22. D. G. Luenberger. *Optimization by Vector Space Methods.* John Wiley and Sons, New York, 1969.
23. J. B. Conway. *A Course in Functional Analysis.* Springer Verlag, New York, 1985.
24. E. Kreysig, *Introductory Functional Analysis with Applications.* John Wiley & Sons, New York 1989.
25. C. A. Desoer, M. Vidyasagar. *Feedback Systems: Input-Output Properties.* Academic Press Inc., New York, 1975.
26. T. Kailath. *Linear Systems.* Prentice-Hall, New Jersey, 1980.
27. S. Barnett. *Matrices in Control Theory.* Robert E. Krieger Publishing Co., 1984.
28. M. Vidyasagar. *Control System Synthesis : a Factorization Approach.* Cambridge, Mass. MIT Press, 1985.
29. N. Elia, M.A. Dahleh. *Controller Design with Multiple Objectives.* IEEE Transactions on Automatic Control, 1997; 42:596-613
30. E. J. Anderson, P. Nash. *Linear Programming in Infinite-Dimensional Spaces.* Wiley & Sons, New York, 1991.
31. J. Hilgert, K.H. Hofmann, J.D. Lawson. *Lie Groups, Convex Cones and Semigroups.* Oxford Science Publications,1989.
32. N. Dunford, J. T. Schwartz. *Linear Operators, Part 1, General Theory.* John Wiley & Sons, New York, 1988.
33. J. S. Freudenberg, D. P. Looze. *Frequency Domain Properties of Scalar and Multivariable Feedback Systems.* Springer-Verlag, Berlin, 1987.
34. X. Chen, J. Wen. *A Linear Matrix Inequality Approach to Discrete-Time Mixed $\ell_1/\mathcal{H}_\infty$-Control Problems.* Proceedings of the 34th IEEE Conference on Decision and Control, 1995; 3670-3675
35. N. Elia, M. A. Dahleh. *ℓ_1 -Minimization with Magnitude Constraints in the Frequency Domain.* Journal of Optimization Theory and Applications, 1997; 93:27-52
36. M. A. Dahleh, Y. Ohta. *A Necessary and Sufficient Condition for Robust BIBO Stability.* Systems & Control Letters 1988; 11:271-275
37. M. Khammash, J. B. Pearson. *Performance Robustness of Discrete-Time Systems with Structured Uncertainty.* IEEE Transactions on Automatic Control, 1991; 36:398-412
38. P.G. Voulgaris. *Optimal $\mathcal{H}_2/\ell_1$ control via duality theory.* IEEE Transactions on Automatic Control, 1995; 40:1881-1888
39. M. V. Salapaka, M. Dahleh, P. G. Voulgaris. *MIMO Optimal Control Design: the Interplay between the $\mathcal{H}_2$ and the ℓ_1 Norms.* Proceedings of the 34th IEEE Conference on Decision and Control, 1995; 4:3682-7, also to appear in IEEE Transaction on Automatic Control.
40. M. V. Salapaka, M. Khammash, M. Dahleh. *Solution of MIMO $\mathcal{H}_2/\ell_1$ problem without zero interpolation.* Proceedings of the 36th IEEE Conference on Decision and Control, 1997; 2:1546-51
41. M. A. Dahleh. *BIBO Stability Robustness for Coprime Factor Perturbations.* IEEE Transactions on Automatic Control, 1992; 37:352-355
42. M. Khammash. *Solution of the ℓ_1 MIMO Control Problem without Zero Interpolation.* Proceedings 33rd Conference on Decision and Control, 1996; 4:040-4045

43. A. E. Barabanov, A. A Sokolov. *Geometrical Approach to the ℓ_1 Optimization Problem*. Proceedings of the 33rd Conference on Decision and Control, 1994; 4:3143-3144
44. F. Blanchini, M. Sznaier. *Persistent Disturbance Rejection via Static Feedback*. IEEE Transactions on Automatic Control, 1995; 40:1127-1131
45. N. Elia, M. A. Dahleh. *Minimization of the Worst-Case Peak to Peak Gain via Dynamic Programming*. Proceedings of American Control Conference, 1997; 3002-3006. Also accepted in IEEE Transaction on Automatic Control.
46. N. Elia, M. A. Dahleh. *A Quadratic Programming Approach for Solving the ℓ_1 Multi-Block Problem*. Proceedings of the 35th IEEE Conference on Decision and Control, 1996; 4:4028-33. Also to appear in IEEE Transactions on Automatic Control, 1998;
47. M. Sznaier, J. Bu. *A Solution to MIMO 4-Block ℓ_1 Optimal Control Problems via Convex Optimization*. Proceedings American Control Conference, 1995; 951-955.
48. K. Zhou. *Robust and Optimal Control*. Prentice-Hall, New Jersey, 1996.
49. I. J. Diaz-Bobillo, M. A. Dahleh. *State Feedback ℓ_1 Optimal Controllers Can be Dynamic*. Systems & Control Letters, 1992; 19:87-93
50. I. J. Fialho, T. T. Georgiu. *ℓ_1 State-feedback Control with a Prescribed Rate of Exponential Convergence*. Proceedings American Control Conference, 1995; 956-960
51. D. P. Bertsekas, I. B. Rhodes. *On the minimax reachability of target sets and target tubes*. Automatica, 1971; 7:233-247
52. A. E. Barabanov, A. A. Sokolov. *Geometrical Solution to ℓ_1-optimization Problem with Combined Conditions*. Proceedings Asian Control Conference, Tokyo Japan, 1994; 3:331-334
53. R. T. Rockfellar. *Convex Analysis*. Princeton University Press, Princeton, 1970.
54. W. Rudin. *Principles of Mathematical Analysis*. McGraw-Hill, New York, 1976.
55. L. Narici, E. Beckenstein. *Topological Vector Spaces*. M. Dekker, New York, 1985.

Lecture Notes in Control and Information Sciences

Edited by M. Thoma

1993–1998 Published Titles:

Vol. 186: Sreenath, N.
Systems Representation of Global Climate Change Models. Foundation for a Systems Science Approach.
288 pp. 1993 [3-540-19824-5]

Vol. 187: Morecki, A.; Bianchi, G.; Jaworeck, K. (Eds)
RoManSy 9: Proceedings of the Ninth CISM-IFToMM Symposium on Theory and Practice of Robots and Manipulators.
476 pp. 1993 [3-540-19834-2]

Vol. 188: Naidu, D. Subbaram
Aeroassisted Orbital Transfer: Guidance and Control Strategies
192 pp. 1993 [3-540-19819-9]

Vol. 189: Ilchmann, A.
Non-Identifier-Based High-Gain Adaptive Control
220 pp. 1993 [3-540-19845-8]

Vol. 190: Chatila, R.; Hirzinger, G. (Eds)
Experimental Robotics II: The 2nd International Symposium, Toulouse, France, June 25-27 1991
580 pp. 1993 [3-540-19851-2]

Vol. 191: Blondel, V.
Simultaneous Stabilization of Linear Systems
212 pp. 1993 [3-540-19862-8]

Vol. 192: Smith, R.S.; Dahleh, M. (Eds)
The Modeling of Uncertainty in Control Systems
412 pp. 1993 [3-540-19870-9]

Vol. 193: Zinober, A.S.I. (Ed.)
Variable Structure and Lyapunov Control
428 pp. 1993 [3-540-19869-5]

Vol. 194: Cao, Xi-Ren
Realization Probabilities: The Dynamics of Queuing Systems
336 pp. 1993 [3-540-19872-5]

Vol. 195: Liu, D.; Michel, A.N.
Dynamical Systems with Saturation Nonlinearities: Analysis and Design
212 pp. 1994 [3-540-19888-1]

Vol. 196: Battilotti, S.
Noninteracting Control with Stability for Nonlinear Systems
196 pp. 1994 [3-540-19891-1]

Vol. 197: Henry, J.; Yvon, J.P. (Eds)
System Modelling and Optimization
975 pp approx. 1994 [3-540-19893-8]

Vol. 198: Winter, H.; Nüßer, H.-G. (Eds)
Advanced Technologies for Air Traffic Flow Management
225 pp approx. 1994 [3-540-19895-4]

Vol. 199: Cohen, G.; Quadrat, J.-P. (Eds)
11th International Conference on Analysis and Optimization of Systems – Discrete Event Systems: Sophia-Antipolis, June 15–16–17, 1994
648 pp. 1994 [3-540-19896-2]

Vol. 200: Yoshikawa, T.; Miyazaki, F. (Eds)
Experimental Robotics III: The 3rd International Symposium, Kyoto, Japan, October 28-30, 1993
624 pp. 1994 [3-540-19905-5]

Vol. 201: Kogan, J.
Robust Stability and Convexity
192 pp. 1994 [3-540-19919-5]

Vol. 202: Francis, B.A.; Tannenbaum, A.R. (Eds)
Feedback Control, Nonlinear Systems, and Complexity
288 pp. 1995 [3-540-19943-8]

Vol. 203: Popkov, Y.S.
Macrosystems Theory and its Applications: Equilibrium Models
344 pp. 1995 [3-540-19955-1]

Vol. 204: Takahashi, S.; Takahara, Y.
Logical Approach to Systems Theory
192 pp. 1995 [3-540-19956-X]

Vol. 205: Kotta, U.
Inversion Method in the Discrete-time Nonlinear Control Systems Synthesis Problems
168 pp. 1995 [3-540-19966-7]

Vol. 206: Aganovic, Z.; Gajic, Z.
Linear Optimal Control of Bilinear Systems with Applications to Singular Perturbations and Weak Coupling
133 pp. 1995 [3-540-19976-4]

Vol. 207: Gabasov, R.; Kirillova, F.M.; Prischepova, S.V.
Optimal Feedback Control
224 pp. 1995 [3-540-19991-8]

Vol. 208: Khalil, H.K.; Chow, J.H.; Ioannou, P.A. (Eds)
Proceedings of Workshop on Advances inControl and its Applications
300 pp. 1995 [3-540-19993-4]

Vol. 209: Foias, C.; Özbay, H.; Tannenbaum, A.
Robust Control of Infinite Dimensional Systems: Frequency Domain Methods
230 pp. 1995 [3-540-19994-2]

Vol. 210: De Wilde, P.
Neural Network Models: An Analysis
164 pp. 1996 [3-540-19995-0]

Vol. 211: Gawronski, W.
Balanced Control of Flexible Structures
280 pp. 1996 [3-540-76017-2]

Vol. 212: Sanchez, A.
Formal Specification and Synthesis of Procedural Controllers for Process Systems
248 pp. 1996 [3-540-76021-0]

Vol. 213: Patra, A.; Rao, G.P.
General Hybrid Orthogonal Functions and their Applications in Systems and Control
144 pp. 1996 [3-540-76039-3]

Vol. 214: Yin, G.; Zhang, Q. (Eds)
Recent Advances in Control and Optimization of Manufacturing Systems
240 pp. 1996 [3-540-76055-5]

Vol. 215: Bonivento, C.; Marro, G.; Zanasi, R. (Eds)
Colloquium on Automatic Control
240 pp. 1996 [3-540-76060-1]

Vol. 216: Kulhavý, R.
Recursive Nonlinear Estimation: A Geometric Approach
244 pp. 1996 [3-540-76063-6]

Vol. 217: Garofalo, F.; Glielmo, L. (Eds)
Robust Control via Variable Structure and Lyapunov Techniques
336 pp. 1996 [3-540-76067-9]

Vol. 218: van der Schaft, A.
L_2 Gain and Passivity Techniques in Nonlinear Control
176 pp. 1996 [3-540-76074-1]

Vol. 219: Berger, M.-O.; Deriche, R.; Herlin, I.; Jaffré, J.; Morel, J.-M. (Eds)
ICAOS '96: 12th International Conference on Analysis and Optimization of Systems - Images, Wavelets and PDEs:
Paris, June 26-28 1996
378 pp. 1996 [3-540-76076-8]

Vol. 220: Brogliato, B.
Nonsmooth Impact Mechanics: Models, Dynamics and Control
420 pp. 1996 [3-540-76079-2]

Vol. 221: Kelkar, A.; Joshi, S.
Control of Nonlinear Multibody Flexible Space Structures
160 pp. 1996 [3-540-76093-8]

Vol. 222: Morse, A.S.
Control Using Logic-Based Switching
288 pp. 1997 [3-540-76097-0]

Vol. 223: Khatib, O.; Salisbury, J.K.
Experimental Robotics IV: The 4th International Symposium, Stanford, California, June 30 - July 2, 1995
596 pp. 1997 [3-540-76133-0]

Vol. 224: Magni, J.-F.; Bennani, S.; Terlouw, J. (Eds)
Robust Flight Control: A Design Challenge
664 pp. 1997 [3-540-76151-9]

Vol. 225: Poznyak, A.S.; Najim, K.
Learning Automata and Stochastic Optimization
219 pp. 1997 [3-540-76154-3]

Vol. 226: Cooperman, G.; Michler, G.; Vinck, H. (Eds)
Workshop on High Performance Computing and Gigabit Local Area Networks
248 pp. 1997 [3-540-76169-1]

Vol. 227: Tarbouriech, S.; Garcia, G. (Eds)
Control of Uncertain Systems with Bounded Inputs
203 pp. 1997 [3-540-76183-7]

Vol. 228: Dugard, L.; Verriest, E.I. (Eds)
Stability and Control of Time-delay Systems
344 pp. 1998 [3-540-76193-4]

Vol. 229: Laumond, J.-P. (Ed.)
Robot Motion Planning and Control
360 pp. 1998 [3-540-76219-1]

Vol. 230: Siciliano, B.; Valavanis, K.P. (Eds)
Control Problems in Robotics and Automation
328 pp. 1998 [3-540-76220-5]

Vol. 231: Emel'yanov, S.V.; Burovoi, I.A.; Levada, F.Yu.
Control of Indefinite Nonlinear Dynamic Systems
196 pp. 1998 [3-540-76245-0]

Vol. 232: Casals, A.; de Almeida, A.T.
Experimental Robotics V: The Fifth International Symposium Barcelona, Catalonia, June 15-18, 1997
190 pp. 1998 [3-540-76218-3]

Vol. 233: Chiacchio, P.; Chiaverini, S.
Complex Robotic Systems
189 pp. 1998 [3-540-76265-5]

Vol. 234: Arena, P., Fortuna, L, Muscato, G and Xibilia, M.G.
Neural Networks in Multidimensional Domains: Fundamentals and New Trends in Modelling and Control
179 pp. 1998 [1-85233-006-6]

Vol. 235: Chen, B.M
H_∞ Control and Its Applications
361 pp. 1998 [1-85233-026-0]

Vol. 236: De Almeida, A.T.; Khatib, O
Autonomous Robotic Systems
283 pp. 1998 [1-85233-036-8]

Vol. 237: Kreigman, D.J., Hagar, G.D. and A.S. Morse
The Confluence of Vision and Control
304 pp. 1998 (1-85233-025-2)